Physical Principles of Wireless Communications

Second Edition

Physical Principles of Wireless Communications

Second Edition

Victor L. Granatstein

CRC Press
Taylor & Francis Group
Boca Raton   London   New York

CRC Press is an imprint of the
Taylor & Francis Group, an **Informa** business

CRC Press
Taylor & Francis Group
6000 Broken Sound Parkway NW, Suite 300
Boca Raton, FL 33487-2742

Version Date: 20120206

International Standard Book Number: 978-1-4398-7897-2 (Hardback)

Library of Congress Cataloging-in-Publication Data

Granatstein, V. L. (Victor L.), 1935-
    Physical principles of wireless communications / Victor L. Granatstein. -- 2nd ed.
        p. cm.
    Includes bibliographical references and index.
    ISBN 978-1-4398-7897-2 (hardback)
    1. Wireless communication systems. I. Title.

TK5103.2.G74 2012
621.384--dc23   2011050847

Visit the Taylor & Francis Web site at
http://www.taylorandfrancis.com

and the CRC Press Web site at
http://www.crcpress.com

I think the history of science gives ample examples that pure investigation has enormous benefit.... And lovely things turn up.

James A. Van Allen, 1999

Many of today's electrical devices (e.g., **radios, generators** and **alternators**) can trace their roots to the basic research conducted by Michael Faraday in 1831. He discovered the **principle of electromagnetic induction,** that is, the relationship between **electricity** and **magnetism.**

From the website,
http://www.lbl.gov/Education/ELSI/
Frames/research-basic-history-f.html

Look on the streets of almost any city in the world, however, and you will see people clutching tiny, pocket computers, better known as mobile phones. Already, even basic handsets have simple web-browsers, calculators and other computing functions. Mobile phones are cheaper, simpler and more reliable than PCs, and market forces—in particular, the combination of pre-paid billing plans and microcredit schemes—are already putting them into the hands of even the world's poorest people. Initiatives to spread PCs in the developing world, in contrast, rely on top-down funding from governments or aid agencies, rather than bottom-up adoption by consumers.

Merchants in Zambia use mobile phones for banking; farmers in Senegal use them to monitor prices; health workers in South Africa use them to update records while visiting patients. All kinds of firms, from giants such as Google to start-ups such as CellBazaar, are working to bring the full benefits of the web to mobile phones. There is no question that the PC has democratised computing and unleashed innovation; but it is the mobile phone that now seems most likely to carry the dream of the "personal computer" to its conclusion.

The Economist, **July 27, 2006**

Dedication

To Batya, Rebecca Miriam (Becky), Abraham Solomon (Solly), Annie Sara Khaya, Leora, Arielle Bella, Aliza Rose, Walter John, and Daniel Jonathan

Contents

List of Figures

List of Tables

Preface to the First Edition

Wireless communications is based on the launching, propagation, and detection of electromagnetic waves usually at radio or microwave frequencies. It has its roots in the middle of the 19th century when James Clerk Maxwell formulated the basic laws of electromagnetism (viz., Maxwell's equations) and Heinrich Hertz demonstrated propagation of radio waves across his laboratory. By the start of the 20th century, Guglielmo Marconi had invented the wireless telegraph and sent signals across the Atlantic Ocean using reflection off the ionosphere. Subsequent early embodiments of wireless communication systems included wireless telephony, AM and FM radio, shortwave radio, television broadcasting, and radar. Engineering breakthroughs after World War II including launching artificial satellites, the miniaturization of electronics, and the invention of electronic computers, led to new embodiments of wireless communication systems that have revolutionized modern lifestyles and created dominant new industries. These include cellular telephones, satellite TV beaming, satellite data transmission, satellite telephones, and wireless networks of computers.

The present textbook presents descriptions of the salient features of these modern wireless communication systems together with rigorous analyses of the devices and physical mechanisms that constitute the physical layers of these systems. Starting with a review of Maxwell's equations, the operation of antennas and antenna arrays is explained in sufficient detail to allow for design calculations. Propagation of electromagnetic waves is also explored leading to useful descriptions of mean path loss through the streets of a city or inside an office building. The principles of probability theory are reviewed so that students will be able to calculate the margins that must be allowed to account for statistical variation in path loss. The physics of geostationary earth orbiting (GEO) satellites and low earth orbiting (LEO) satellites are covered in sufficient detail to evaluate and make first-order designs of satellite communications (SATCOM) systems.

This textbook is the outgrowth of a course in the physics of wireless communications that I have taught to electrical engineering seniors and first-year graduate students at the University of Maryland for the past 7 years. I have also been invited by Tel Aviv University (TAU) to present an accelerated version of the course to graduate students and working engineers at wireless communication companies; I have presented such a course at TAU on two occasions, in 2003 and 2004 to 2005. The course at the University of Maryland is a senior elective course that is normally limited to 30 students, but because of its popularity the class size was expanded to as many as 60 students. Problem sets have also been developed and are included; a solutions manual is available for instructors.

Previous textbooks have tended to be of two types as follows:

1. Those that stress systems and signal processing aspects of wireless communications with relatively light treatment of antennas and propagation
2. Those that stress antennas and propagation with little attention paid to the details of modern communication systems

The present textbook aims to integrate the topical area of antennas and propagation with consideration of its application to designing the physical layer in modern communication systems. This textbook aims to provide the following:

1. Historical treatment of wireless communications from Marconi's wireless telegraph to today's multimedia wideband transmissions
2. Starting from Maxwell's equations, to analyze antennas and propagation as they relate to modern communication systems
3. Relevant treatments of noise and statistical analysis
4. Integration of electromagnetic analysis with complete descriptions of the physical layer in the most important wireless systems including cellular/PCS telephones, wireless local area networks of computers, and GEO and LEO SATCOM

Victor Granatstein

Silver Spring, Maryland

January 2007

Preface to the Second Edition

The technology of wireless communications is changing rapidly. In the less than 5 years that have elapsed since the writing of the first edition of this book, smartphones with Internet access have become ubiquitous, making it appropriate to add material in the first chapter describing third-generation (3G) and fourth-generation (4G) cell phone systems.

Beyond that natural update, I have received many suggestions for adding new material on topics that were treated lightly or omitted in the first edition. Primary among these were discussions of the relation between the basic laws of quantum and relativistic physics and the engineering of modern wireless communication systems.

A section has been added describing Planck's law of blackbody radiation. This has been followed by a detailed description of the assumptions made to derive from this law the engineering estimates of noise pickup by a communications receiver.

Discussions of both general relativity and special relativity have been expanded in the context of synchronizing clocks on the earth and in a satellite. A global positioning system (GPS) cannot be successfully designed without taking relativistic effects into account.

The chapter on antennas has been made more complete by adding a section on wire loop antennas. In the chapter on statistical design, the discussion of shadowing correlations and their effect on cell phone system design has been expanded.

The second edition will make clearer to students the relationship between discoveries in pure science and their application to the invention and engineering of wireless communication systems that are such an important component in shaping the world of the 21st century. Understanding the inspiring efforts of the scientists and engineers who have contributed to the communications revolution will hopefully lead a new generation of innovators to pave the way for as yet unimagined marvels.

Victor L. Granatstein

Silver Spring, Maryland

August 2011

The Author

Victor L. Granatstein was born and raised in Toronto, Canada. He received the Ph.D. degree in electrical engineering from Columbia University, New York, in 1963. After a year of post-doctoral work at Columbia, he became a research scientist at Bell Telephone Laboratories from 1964 to 1972 where he studied microwave scattering from turbulent plasma. In 1972, he joined the Naval Research Laboratory (NRL) as a research physicist, and from 1978 to 1983, he served as head of NRL's High Power Electromagnetic Radiation Branch.

In August 1983, he became a professor in the Electrical Engineering Department of the University of Maryland, College Park. From 1988 to 1998, he was director of the Institute for Plasma Research at the University of Maryland. Since 2008, he has been director of research of the Center for Applied Electromagnetics at the University of Maryland. His research has involved invention and development of high-power microwave sources for heating plasmas in controlled thermo-nuclear fusion experiments, for driving electron accelerators used in high-energy physics research, and for radar systems with advanced capabilities. He also has led studies of the effects of high-power microwaves on integrated electronics. His most recent study is of air breakdown in the presence of both terahertz radiation and gamma rays with possible application to detecting concealed radioactive material. He has coauthored more than 250 research papers in scientific journals and has coedited three books. He holds a number of patents on active and passive microwave devices.

Granatstein is a fellow of the American Physical Society (APS) and a life fellow of the Institute of Electrical and Electronic Engineers (IEEE). He has received a number of major research awards including the E.O. Hulbert Annual Science Award (1979), the Superior Civilian Service Award (1980), the Captain Robert Dexter Conrad Award for scientific achievement (awarded by the Secretary of the Navy, 1981), the IEEE Plasma Science and Applications Award (1991), and the Robert L. Woods Award for Excellence in Electronics Technology (1998). He has spent part of his sabbaticals in 1994, 2003, and 2010 at Tel Aviv University where he holds the position of Sackler Professor by Special Appointment.

He lives in Silver Spring, Maryland, with his wife Batya; they recently celebrated their 56th wedding anniversary. They have three children, Rebecca, Solly, and Annie, and to date, three grandchildren, Leora, Arielle Bella, and Aliza Rose.

Acknowledgments

The author is indebted to his students, especially Ioannis Stamatiou who ably assisted in preparing the figures, and to Ankur Jain who proofread the entire first edition manuscript. (Ioannis Stamatiou received his Master of Science in Electrical Engineering in December 2006. Ankur Jain received his Bachelor of Science in Electrical Engineering in August 2006.) The author is also grateful to his department chairman, Patrick O' Shea, who generously provided resources and encouragement in support of preparing this manuscript. Professor Avraham Gover who used the first edition in classes at Tel Aviv University made a number of very useful suggestions that are incorporated into the second edition.

CHAPTER 1

An Introduction to Modern Wireless Communications

1.1 A Brief History of Wireless Communications

Before 1844, long-distance communications depended on physically transporting messages on horseback (e.g., the Pony Express) or in ocean-crossing vessels. Weeks might pass before Americans learned of important events in Europe. This situation was radically altered by the advent of the electric telegraph that enabled almost instantaneous message transmission. During the course of the 19th century, telegraph wires were strung across continents, and a cable was laid across the floor of the Atlantic Ocean. However, there was no possibility of using this technology to communicate with people in motion (e.g., on oceangoing ships). Also, in the poorer or more sparsely settled regions of the world it was not economically feasible to string wires over long distances. Thus, the discovery of radio frequency (RF) waves and their deployment in wireless communications, as the 19th century changed into the 20th, constituted a second communications revolution. Wireless communications was a liberating technology, untethering transmitters and receivers from wires or cables and making communication with mobile receivers practical. Moreover, wireless communications was an equalizing technology empowering vast regions of the less-developed part of the world with a means of virtually instantaneous, long-distance communications.

To appreciate the impact of communications using RF waves, one has only to recall that before 1895, the British Admiralty, employing the best available technology, communicated with its fleet in the English Channel by using telescopes, flags, and

1

flashing lights. This visual communication system functioned over that limited range as long as the weather was clear, but it was defeated by fog, which unfortunately was a common occurrence in the English Channel. Contrast that with the situation a century later when RF waves traveling through space without wires or cables carry voice, data, and picture messages between points on Earth separated by thousands of miles in any type of weather and even carry commands from Earth to spacecraft at the outer edges of the solar system.

1.1.1 Faraday, Maxwell, and Hertz: The Discovery of Electromagnetic Waves

Wireless RF communications is truly a marvel. It represents an outstanding achievement of the scientific method and of science-based engineering. For many centuries, it had been appreciated that magnets, the first instances of which were naturally occurring "magic stones," could exert a force on each other and on certain metals *at a distance*; the agent for transmitting this force was called a magnetic field. Similarly, electrically charged bodies produced in the simplest case by friction (e.g., rubbing a glass rod with a silk cloth) could also exert a force on each other *at a distance*; the agent for transmitting this force was called the electric field. By 1831, the great experimental physicist, Michael Faraday (1791 to 1867, b. London, England) had discovered that the two types of fields were related; Faraday's Law of electromagnetic induction stipulated that *an electric field could be generated by a time-varying magnetic field*. This can be written mathematically in point form and for the fields in free space as

$$\nabla \times \mathbf{E}\,(\mathbf{r},t) = -\mu_o\, \partial \mathbf{H}(\mathbf{r},t)/\partial t \qquad (1.1)$$

where $\mathbf{E}(\mathbf{r},t)$ is the electric field, a function of the spatial coordinates $\mathbf{r}$ and time t; $\mathbf{H}(\mathbf{r},t)$ is the magnetic field; and, the permeability of free space in meter-kilogram-second-ampere (MKSA) units is given by $\mu_o = 4\pi \times 10^{-7}$ Henries/meter. (MKSA units will be used throughout this text. Boldface type denotes vector quantities.)

Faraday's law was the basis for developing powerful electric generators leading to the electrification of factories and whole cities in the 19th and early 20th centuries.

Even though he was a brilliant experimentalist, Faraday realized his limitations and appealed to the theoretician, James Clerk Maxwell (1831 to 1879, b. Edinburgh, Scotland) to develop an exact mathematical description of all the known attributes of electromagnetic phenomena. Of momentous significance was Maxwell's postulate that because nature is often symmetric and reciprocal, as a complement to Faraday's law, *a magnetic field could be generated by a time-varying electric field*. This was stated mathematically in 1864 (again for free space) by

$$\nabla \times \mathbf{H}\ (\mathbf{r},\ t) = \varepsilon_o\ \partial \mathbf{E}(\mathbf{r},\ t)/\partial t \qquad (1.2)$$

where the permittivity of free space is a constant given by

$$\varepsilon_o = 8.854 \times 10^{-12}\ \text{Farads/meter}$$

$$\approx (36\ \pi \times 10^9)^{-1}\ \text{Farads/meter}$$

It is straightforward to combine Equations (1.1) and (1.2) to obtain a second-order equation in either $\mathbf{E}$ or $\mathbf{H}$ of the form

$$\nabla^2 \mathbf{E}\ (\mathbf{r},t) - c^{-2}\ \partial^2 \mathbf{E}(\mathbf{r},t)/\partial t^2 = 0 \qquad (1.3)$$

where $c = (\varepsilon_o \mu_o)^{-\frac{1}{2}} = 2.998 \times 10^8$ meters/second.

The solutions of Equation (1.3) are especially simple in form for the well-known plane wave case (e.g., electric field linearly polarized in the x direction with spatial variation only in the z direction or $\mathbf{E}(\mathbf{r},t) = E(z,t)\mathbf{a}_x$. In that case the solutions of Equation (1.3) which also, of course, satisfy Equations (1.1) and (1.2) are

$$\mathbf{E}(\mathbf{r},t) = E_1 \cos\ (\omega[t - z/c] + \chi_1)\ \mathbf{a}_x$$

and

$$\mathbf{H}(\mathbf{r},t) = (E_1/Z_o)\ \cos\ (\omega[t - z/c] + \chi_1)\mathbf{a}_y \qquad (1.4a)$$

or

$$\mathbf{E}(\mathbf{r},t) = E_2 \cos\ (\omega[t + z/c] + \chi_2)\mathbf{a}_x$$

and

$$\mathbf{H}(\mathbf{r},t) = -(E_2/Z_o)\ \cos\ (\omega[t + z/c] + \chi_2)\mathbf{a}_y \qquad (1.4b)$$

where the free space wave impedance $Z_o = (\mu_o/\varepsilon_o)^{1/2} \approx 120\pi$ Ohms = 377 Ohms; $E_{1,2}$ and $\chi_{1,2}$ are, respectively, electric field amplitudes and phases, and ω is the wave angular frequency (frequency in Hertz, f = w/2p).

The solutions given by Equation (1.4) have a number of remarkable properties. They describe waves of coupled electric and magnetic fields propagating through free space in either the positive z (Equation 1.4a) or negative z directions (Equation 1.4b). The electric and magnetic fields are orthogonal to each other, have the same phase, and propagate through space at a speed of $c = 2.998 \times 10^8$ meters per second (i.e., 186,000 miles per second) in a direction given by $\mathbf{E} \times \mathbf{H}$. This speed c is also the speed of visible light, which had been measured with reasonable accuracy as early as the 17th century, strongly suggesting to Maxwell that light was composed of electromagnetic waves although in a restricted part of the frequency spectrum (viz., 4×10^{14} Hz < f < 7.5×10^{14} Hz or $0.4\ \mu m < \lambda < 0.8\ \mu m$ where the wavelength $\lambda = c / f$).

Equation (1.4) predicts the existence of electromagnetic waves unrestricted in frequency. Radio frequency (RF) waves is a term that will be used to designate electromagnetic waves over the frequency range 30 kHz to 300 GHz encompassing *radio waves* (30 kHz to 300 MHz) and *microwaves* (300 MHz to 300 GHz) with the subset of microwaves from 30 GHz to 300 GHz being designated *millimeter waves*. Finer subdivisions of the RF spectral range are displayed in Table 1.1. Modern wireless communication systems with which this book is concerned utilize microwaves and millimeter waves.

The microwave ultra high frequency (UHF) and super high frequency (SHF) and millimeter wave extremely high frequency (EHF) parts of the spectrum are frequently subdivided into still smaller portions corresponding to standard waveguide sizes with the frequency bands designated by letters. The microwave and millimeter-wave frequency bands as defined by the Radio Society of Great Britain are displayed in Table 1.2.

Heinrich Hertz (1857 to 1894, b. Hamburg, Germany) was the major interpreter of the implications of Maxwell's theory of electromagnetism. Within 24 years of Maxwell's prediction of the existence of RF waves, Hertz had experimentally verified their existence and shown that their basic properties were the same as light waves. He generated the waves at a frequency near 100 MHz (wavelength 1 = c/f = 3 m) using a resonant circuit. The waves then propagated across his laboratory and produced a sparking in

TABLE 1.1 The Radio Frequency (RF) Spectrum

Band Designation	Communications Application	Frequency	Wavelength
SLF, super-low frequency	Submarine communications	30–300 Hz	100–1000 km
ULF, ultra-low frequency	Audio signal modulation	0.3–3 kHz	10–100 km
VLF, very low frequency	Navigation and position location	3–30 kHz	1–10 km
LF, low frequency	Weather broadcast stations	30–300 kHz	0.1–1 km
MF, medium frequency	AM radio, "ground wave"	0.3–3 MHz	10–100 m
HF, high frequency	Shortwave radio, "sky wave"	3–30 MHz	1–10 m
VHF, very high frequency	FM radio, TV, mobile radio, air traffic control	30–300 MHz	0.1–1 m
UHF, ultra-high frequency	UHF-TV, cellular phones, local area networks (LANs), SATCOM, global positioning systems (GPSs)	0.3–3 GHz	10–100 cm
SHF, super-high frequency	Radar, SATCOM	3–30 GHz	1–10 cm
EHF, extremely high frequency	Military radar and SATCOM	30–300 GHz	1–10 mm
Submillimeter		0.3–3 THz	0.1–1 mm
Far infrared		3–30 THz	10–100 μm

Note: kHz = 10^3 Hz, MHz = 10^6 Hz, GHz = 10^9 Hz, THz = 10^{12} Hz.

a small gap at the center of a metal rod (a dipole antenna). There were no connecting wires between the receiver and the generator. By reflecting the waves from metal plates, Hertz was able to set up standing waves and determine the wavelength from the distance between nulls in the standing wave pattern. In addition to reflection, he demonstrated that, like light, the radio waves could be refracted and polarized. Hertz was able to declare "Optics is no longer restricted to (visible light) waves, a small fraction of a millimeter in (wave)length; its domain is extended to wave(length)s that are measured in decimeters, meters and kilometers."[1]

---

[1] L. S. Lawrence, *Physics for Scientists and Engineers*, Volume 2 (Jones and Bartlett Publishers, London, U.K. 1996) p. 929.

TABLE 1.2   Microwave and Millimeter-Wave Waveguide Bands

Band Designation	Frequency Range
L-band	1–2 GHz
S-band	2–4 GHz
C-band	4–8 GHz
X-band	8–12 GHz
Ku-band	12–18 GHz
K-band	18–26.5 GHz
Ka-band	26.5–40 GHz
Q-band	30–50 GHz
U-band	40–60 GHz
V-band	50–75 GHz
E-band	60–90 GHz
W-band	75–110 GHz
F-band	90–140 GHz
D-band	110–170 GHz

1.1.2 Guglielmo Marconi, Inventor of Wireless Communications

While Maxwell and Hertz were practitioners of basic scientific research engaged in uncovering and understanding new and unexpected physical phenomena, their work led directly to the invention of wireless communications. Credit for this invention is a matter of some controversy, with credit having been claimed both by Nikolai Tesla (1865 to 1943) and by Guglielmo Marconi (1874 to 1937, b. Bologna, Italy). Marconi filed for a patent on the wireless telegraph in 1896, and the following year Tesla filed for a patent with improved circuit features. Nevertheless, the major credit for the practical development of the wireless telegraph certainly belongs to Marconi.

As a teenager, Marconi had studied the experiments of Hertz with keen interest. He began a systematic effort of increasing the power of the radio wave transmission and extending its range. He concentrated on improving the sensitivity of the receiver and on increasing the size of the antenna; at some early stage he began using the oscillator circuits devised by Tesla. By 1895, when Marconi was only 21 years old, he had demonstrated radio wave propagation and detection over distances larger than a mile. With the help of business contacts of his Irish mother, he had established a company in England that was based in part on successfully convincing the British Admiralty that they might overcome their fog problem by communicating shore-to-ship in the English Channel using wire-less telegraphy. In 1897, he transmitted signals from shore to a ship 18 miles away. By 1899, his company had established a commercial wireless telegraph link from England to France that operated over a distance of 85 miles in all kinds of weather.

By the end of 1901, Marconi attempted a much more ambitious, if rather improbable, experiment trying to communicate with radio waves across the Atlantic Ocean from Poldhu, Cornwall, United Kingdom, to St. Johns, Newfoundland. There were many knowl-edgeable scientists who were skeptical regarding the chances of success because the curvature of the earth would prevent the wave from traveling in a straight line over such a long distance; how-ever, to almost everyone's surprise, the experiment was successful. The letter "s," three dots in Morse code, was transmitted from at preset intervals across the Atlantic on December 12, 1901, and detected at Marconi's station in Newfoundland. This caused some degree of panic in the cable telegraph company that had spent a considerable fortune in laying an underwater trans-Atlantic cable. Their staff at the cable station in North Sydney, Nova Scotia, Canada, sent this poetic transmission to their colleagues in the cable office in Liverpool England shortly after Marconi's success:

> Best Christmas greetings from North Sydney
> Hope you are sound in heart and kidney,
> Next year will find us quite unable
> To send exchanges o'er the cable:
> Marconi will our finish see,
> The cable co's have ceased to be;
> No further need of automatics
> Retards, resistances and statics.
> I'll then across the ether sea
> Waft Christmas greetings unto thee.[2]

This little poem proved to be quite prescient in that full-text messages were being sent across the Atlantic by the following December in spite of the fact that Marconi had been forced to leave Newfoundland and relocate his North American station to Nova Scotia because of a threatened lawsuit by the cable telegraph company thta held a monopoly in Newfoundland. The predicted demise of the cable company was however a bit too pessimistic. The cable companies and Marconi's wireless telegraph company

---

[2] D. M. McNicol, *Radio's Conquest of Space* (J. J. Little and Ives Company, New York, 1946) p. 142

competed vigorously in the early 20th century much like today's competition between cable and satellite-based TV programming providers. Wireless telegraphy however did have the unique advantage of being able to serve mobile customers such as ocean-going shipping, a precursor of today's cell phone used by customers who are walking or riding in automobiles.

By the end of 1902, Arthur Kennelly in the United States and Oliver Heaviside in England correctly postulated that Marconi's trans-Atlantic radio waves were bouncing off an electrically charged region several hundred kilometers above the Earth's surface. This region was at first called the Kennelly–Heaviside layer, and later, when its structure was understood to be a complex of many layers of ionized air molecules, it was called the ionosphere. The existence of the ionosphere was not even guessed at before Marconi's experiments in 1901 and 1902. Thus, experiments by an inventor and entrepreneur that were investigating the possibility of an intercontinental wireless telegraph business, led to the discovery of the ionosphere, one of the salient features of planet Earth. It should be noted that it was fortuitous that Marconi had altered Hertz's apparatus in attempting to increase the range between transmitter and receiver because his changes also lowered the RF frequency. The 100 MHz signal of Hertz would not have reflected off the ionosphere; Marconi's signal, which was tuned for maximum power at 850 kHz, probably had a component at ~7 MHz that could be reflected with sufficient range even in the daytime transmission of December 1901. By December 1902, Marconi had realized that the reflection of a 850 kHz signal was more effective at night; this is because the nighttime ionosphere is less dense and the position of the layer for reflecting 850 kHz signals lies at an appropriate altitude for trans-Atlantic reflection. In 1909, Marconi was awarded the Nobel Prize in Physics for his work on wireless telegraphy, the progenitor of all wireless communications.

Pictures of the scientists who laid the foundation for wireless communications, Michael Faraday, James Clerk Maxwell, Heinrich Hertz, and Guglielmo Marconi, are displayed in Figure 1.1.

1.1.3 Developments in the Vacuum Electronics Era (1906 to 1947)

In 1893, Thomas Edison had observed that electric current could be passed unidirectionally through a vacuum tube between a heated filament and a plate that was at a positive voltage with respect to the filament, and this was subsequently understood as being due to

(a)

(b)

FIGURE 1.1   (a) Michael Faraday, discoverer of the relationship between electricity and magnetism. (b) James Clerk Maxwell in about 1857 (age 26), the year in which he first wrote to Michael Faraday.

(c)

(d)

FIGURE 1.1 (CONTINUED) (c) Heinrich Hertz, the discoverer of radio waves. (d) Guglielmo Marconi, with an early version of his wireless telegraph on which he filed a patent in 1896 at age 22.

the flow of electrons emitted by the cathode filament through the vacuum to the anode plate. In 1904, J.A. Fleming demonstrated that Edison's vacuum diode was useful as a detector of radio waves. This was followed in 1906 by Lee DeForest's momentous invention of the triode vacuum amplifier in which a small RF signal applied between the cathode and a grid resulted in an amplified RF signal in the anode-cathode circuit. The advent of RF amplifiers greatly enhanced the capability of radio communications. Following closely after DeForest's invention, Reginald Fessenden demonstrated radio telephony, the transmission of voice signals on an RF carrier. Radio telephony developed rapidly during World War I. By 1916, David Sarnoff, who had been a Marconi employee, proposed commercial radio broadcasting with radio receivers ("radio music boxes") becoming household appliances. Commercial radio broadcasting became a reality in 1920 when station KDKA in Pittsburgh, Pennsylvania, broadcast the ongoing vote tally in the Harding–Cox U.S. presidential election. Also in the early 1920s, U.S. police cars began using mobile radio communications, although the bulky and heavy vacuum electronic transmitters and receivers must have occupied most of their luggage space.

Other important inventions based on RF waves followed on the heels of wireless telegraphy, wireless telephony, and radio broadcasting. By 1927, Philo Farnsworth had patented television, and in 1931, Edwin Armstrong patented FM radio. One of the most important advances was the development of high-power vacuum tubes that could generate signals at microwave frequencies. In 1939, J.T. Randall and H.A.H. Boot developed a high-power magnetron at 3000 MHz for use in radar capable of long-range detection of German bombers approaching Britain; this has been cited as the most important scientific contribution to the defeat of the axis powers in World War II.

1.1.4 The Modern Era in Wireless Communications (1947 to the Present)

The two events that enabled the modern era in wireless communications were the launching of artificial earth satellites and the invention of the transistor leading to the miniaturization of electronics. Arthur C. Clarke proposed geostationary communications satellites in 1945 at the end of World War II. The first satellite, Sputnik 1, was launched by the USSR on October 4, 1957, and was equipped with RF transmitters (radio beacons) operating at frequencies of 20 MHz and 40 MHz. Playing catch-up, the United States launched its first satellite, Explorer 1, on January 31, 1958. The first communications satellite with an active transponder was Telstar 1, which transmitted a short television program from Andover, Maine, to Pleumeur-Bodou, France, on July 10, 1962. Telstar 1 also transmitted phone calls, radio programs, and newspaper articles. Telstar was conceived by J.R. Pierce and his colleagues at Bell Telephone Laboratories.

On July 11, 1962, President John F. Kennedy released the following statement on the Telstar achievement:

> The achievement of the communications satellite, while only a prelude already throws open to us the vision of an era of international communications. There is no more important field at the present time than communications and we must grasp the advantages presented to us by the communications satellite and use this medium wisely and effectively to insure greater understanding among the peoples of the world.[3]

Communication satellites removed the upper frequency bound on long-distance wireless communications imposed by the physics of ionospheric reflection (about 60 MHz). Microwave and even millimeter wave signals with their much larger information carrying capacity could now be "bounced" around the world.

The active transponder in Telstar 1 and in subsequent communications satellites used transistors, solid-state RF amplifiers, which had been invented a few years before Telstar 1 was launched. Because of their lighter weight and smaller size, transistors were more suitable for space deployment than vacuum tube RF amplifiers.

The transistor had been invented at Bell Telephone Laboratories in 1947 by John Bardeen, Walter Brattain, and William Shockley, an achievement for which they were awarded the Nobel Prize in Physics in 1956. It was based on the control of electron flow in semiconductor materials and required completely different fabrication techniques than vacuum tube amplifiers. In 1958, Jack Kilby of Texas Instruments invented the integrated circuit in which a circuit containing many transistors was fabricated on a single microchip. Improvements in fabrication techniques for microchips have resulted in increasing miniaturization so that the number of transistors on a chip has been doubling every 1.5 years (Moore's Law); Intel's Tukwila chip contained more than 2 billion transistors in 2008.

A major beneficiary of the miniaturization of electronics was wireless mobile communications, because wireless telephones and even computers with wireless access to the Internet could be made small enough to fit into a purse or a shirt pocket. To fully exploit the potential mass market for smaller (and cheaper) mobile telephones, new strategies had to be devised for sharing the available

---

[3] J. K. Kennedy Press Conference, July 23, 1962, Boston MA: John F, Kennedy Library and Museum; available from http://www.jfklibrary.org/jfkpressconference620723.html

bandwidth in the electromagnetic spectrum among an increasing number of customers. In 1969, Bell Laboratories devised the cellular telephone scheme in which a service area was divided into a matrix of cells, and frequency channels were reused in nonadjacent cells. The first cell phone system to begin operations was Nippon Telegraph and Telephone Corporation (NTT's) in Japan in 1979. The first-generation U.S. cell phone system (1G) known by the acronym of AMPS (Advanced Mobile Phone System) began operating in 1983 when the Federal Communications Commission (FCC) allotted frequencies from 824 MHz to 894 MHz for cellular telephone use. The Global System for Mobile Communications (GSM) digital cellular system began operations in Europe in 1991. In the United States and Japan, the second generation of cell phones (2G) has also used digital modulation. The third generation of cell phones (3G) provides such advanced features as always-on Internet access. The number of mobile communication subscribers has been increasing at a phenomenal rate from 25,000 in 1984 to 5 billion in 2010. The amount of information transmitted per subscriber is also rapidly increasing from thousands of bits per month per subscriber in the 1960s (mostly paging alerts) to millions of bits per month per subscriber projected as multimedia transmissions become ubiquitous. The number of subscribers will of course saturate because the earth's population is finite, but for the foreseeable future because of increasing message content, message traffic is expected to keep growing at the rate of two orders of magnitude per decade.

Before ending this historical account and beginning an overview of basic concepts, a short note on Bell Telephone Laboratories may be in order. As described above, Bell Laboratories was responsible for inventions that have revolutionized electronic technology during the last 50 years (e.g., the transistor, the communications satellite, the cellular telephone concept, the laser). Since winning the Nobel Prize in 1956 for the invention of the transistor, researchers at Bell Laboratories were awarded four additional Nobel prizes in physics. Bell Laboratories also has been awarded over 30,000 patents. This spectacularly successful research organization was the result of the U.S. government allowing the AT&T regulated telephone monopoly to use a share of its profits for the support of an academic research organization that could attract the most able researchers by paying them salaries above the university norm and allowing them to perform research without responsibility for teaching or raising funding for their investigations. In 1984, the U.S. government broke up the AT&T monopoly requiring the divestiture of the local telephone companies and opening long-distance calling to competition. Quite predictably, within 10 years, the downsizing and conversion of Bell laboratories to a business-oriented industrial development laboratory began in earnest. It was an oversight of stunning destructiveness that no measures were taken to protect the world's leading academic research at Bell Laboratories when the 1984 divestiture took place.

1.2 Basic Concepts

1.2.1 Information Capacity of a Communication Channel

Central to the performance of any communications channel is the rate at which it can transfer information. This is specified by the channel information capacity, C_I, which is measured in bits per second. The information theory of Claude Shannon (1916 to 2001, b. Petoskey, Michigan), developed at Bell Telephone Laboratories, relates channel information capacity to frequency bandwidth of the channel and the signal-to-noise ratio at the receiver. According to Shannon, the maximum value of channel information capacity is given by

$$C_I = \Delta f \log_2 [1 + (P_s/P_N)] \tag{1.5}$$

where Δf is the channel bandwidth in Hz, P_s is the signal power at the receiver input, and P_N is the equivalent noise power at the receiver input.

Equivalent noise power accounts for both noise delivered to the receiver input from external sources and noise generated in the receiver itself.

It is seen from the form of Equation (1.5) that the parameters on which C_I depends are noise power, signal power, and bandwidth. We will be concerned in this book with calculating these parameters for modern communication systems operating in various regions of the RF spectrum. As requirements for high-speed data transmission, a premium is placed on having more available bandwidth. This has pushed modern communications systems to higher operating frequencies in the microwave and millimeter wave regions; the millimeter wave band (30–300 GHz) spans a bandwidth of 270 GHz, a factor of nine more bandwidth than all the lower-frequency RF bands.

1.2.2 Antenna Fundamentals

Undoubtedly, we are all aware of antennas attached to RF transmitters and receivers (e.g., those little rods sticking up from our cell phones or the dish-like structures on rooftops for receiving TV programming from satellites). One uses a transmitting antenna to focus or direct the transmitter power toward the intended receiver or receivers. The least directive (worst) antenna that one can imagine is an isotropic radiator that would produce a power density at a distance, d, from a transmitter with radiated power, P_r, of

$$S_I = P_r/(4\pi \, d^2) \tag{1.6}$$

A real antenna is characterized by its directivity, D, which is the ratio of the maximum power density that could be radiated by the antenna in a preferred direction in free space divided by S_I—that is,

$$D = S_{max}/S_I \tag{1.7}$$

Power density is increased by focusing the radiation beam in space. Thus, one can speak of the width of the beam of radiation produced by an antenna; the angular extent of the region where power density in the beam is at least one half of the maximum power density is called the 3 dB beamwidth. Beamwidth might be expected to vary inversely with the directivity. For an isotropic radiator with wave propagation in the outward radial direction in spherical coordinates, the product of the 3 dB beamwidth in the azimuthal direction, BW_ϕ, and the 3 dB beamwidth in the direction of the polar angle, BW_θ, is approximately equal to 4π steradians—that is,

$$BW_\phi \, BW_\theta \approx 4\pi \tag{1.8}$$

An antenna with directivity $D > 1$ will narrow the beamwidths as

$$BW_\phi \, BW_\theta \approx 4\pi/D \tag{1.9}$$

or inversely,

$$D \approx 4\pi/[BW_\phi \, BW_\theta] \tag{1.10}$$

If each beamwidth were expressed in degrees instead of radians, Equation (1.10) can be written

$$D \approx 41,000 \ (\text{deg.}^2)/[\text{BW}_\phi(\text{deg.}) \ \text{BW}_\theta(\text{deg.})] \qquad (1.11)$$

For the important case of an antenna that is omnidirectional in the horizontal plane, $\text{BW}_\phi(\text{deg.}) = 360°$ and Equation (1.11) becomes

$$D \approx 114°/[\text{BW}_\theta(\text{deg.})] \qquad (1.12)$$

More exactly for an omnidirectional antenna,

$$D = 102°/[\text{BW}_\theta(\text{deg.}) - 0.0027 \ \{\text{BW}_\theta(\text{deg.})\}^2] \qquad (1.13)$$

Typically, a real antenna will consist of a configuration of conductors in which RF currents flow resulting in radiation. There will be Ohmic losses in the conductors that one tries to keep small. The antenna efficiency, η, is the ratio of the radiated power to the total power fed into the antenna. Thus, one can define antenna gain, G, which is a measure of antenna effectiveness in producing power density in a preferred direction, as

$$G = \eta D \qquad (1.14)$$

The receiver will also have an antenna attached to collect the incoming radiation. Frequently, for a device such as a cell phone, which contains both a transmitter and a receiver, a single antenna is used for both receiving and transmitting. It will be shown in Chapter 3 that the effective area presented by a receiving antenna when collecting radiation is proportional to the gain, G, that it would have as a transmitting antenna; namely,

$$G = (4\pi/\lambda^2)A_e \qquad (1.15)$$

where A_e is the effective area of the antenna.

For large aperture antennas such as parabolic reflectors or waveguide horns, the effective area is approximately equal to (but a little smaller than) the physical area of the aperture.

1.2.3 The Basic Layout of a Wireless Communications System

A schematic representation of a basic one-way wireless communication link is shown in Figure 1.2. The transmitter produces power at the RF operating frequency, which is modulated by an analog or digital signal. The transmitter is characterized by its output power, P_t, in Watts. A feeder cable connecting the transmitter to

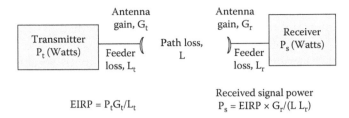

EIRP = P_tG_t/L_t

Received signal power
P_s = EIRP × $G_r/(L\ L_r)$

FIGURE 1.2 Basic arrangement for a one-way wireless communication system. Antennas interface between the transmitter (or receiver) circuit and the wave propagation medium (e.g., free space, space filled with obstacles such as tall buildings, etc.).

its antenna is characterized by its loss, L_t, which is a dimensionless ratio of power at the cable input divided by power at the cable output. The transmitting antenna provides the interface between the transmitter circuit and the wave propagation medium and is characterized by its gain, G_t. The overall capability of the transmitter, the transmitter cable, and the transmitting antenna is often characterized by a parameter called the effective isotropic radiated power (EIRP) given by

$$\text{EIRP} = P_tG_t/L_t \qquad (1.16)$$

The wave propagation medium may be simply free space, but often it is complicated by the presence of physical objects that may absorb, reflect, refract, or diffract the electromagnetic wave (e.g., atmospheric gases, rain, the ground, buildings). The wave propagation medium is characterized by the path loss, L, which is dimensionless like other loss factors. L is calculated taking into account the decrease in power density due to the range, d, and all the significant effects of the gases and objects along the path or paths between the transmitting and receiving antennas; phase interference between waves reaching the receiving antenna along different paths will also influence the path loss.

The receiving antenna is characterized by its gain, G_r, which is a measure of its effectiveness in collecting power to be delivered to the receiver. This collected power will be attenuated by the losses in the cable connecting the receiving antenna to the receiver, L_r. The relationship between path loss and the RF power delivered to the receiver signal input power, P_s, is defined by

$$P_s = \frac{P_t\ G_t\ G_r}{L_t\ L_r\ L} = \frac{\text{EIRP}\ G_r}{L_r\ L} \qquad (1.17)$$

When component parameters, such as EIRP, G_r, and L_r, are known and the minimum value of P_s is specified in order to produce acceptable signal reception in the presence of noise, Equation (1.17) can be used to determine the maximum acceptable path loss, L_{max}, implying a maximum range. One usually makes such link budget calculations using decibels.

For two-way communications, both the transmitter and the receiver in Figure 1.2 would be replaced by combined transmitter/receiver units (T/R units). For example, both a transmitter and a receiver are found in a cell phone. Usually, each antenna and feeder cable does double duty being used both for transmitting signals and for receiving signals.

1.2.4 Decibels and Link Budgets

Decibel (dB) notation is frequently used in making power calculations. The power in dB is obtained by taking the logarithm of the dimensionless ratio of the power, P, to a reference value of power, P_{ref}, and multiplying the result by 10:

$$P(dB) = 10 \log [P/P_{ref}] \tag{1.18}$$

If the reference power is 1 Watt, the letter W is added to the end of dB to indicate that fact:

$$P(dBW) = 10 \log[P/(1 \text{ Watt})] \tag{1.19}$$

If the reference power is 1 milliwatt, the letter m is added:

$$P(dBm) = 10 \log [P/(1 \text{ milliwatt})] \tag{1.20}$$

Note that

$$P(dBW) = P(dBm) - 30 \tag{1.21}$$

The inverse of Equation (1.18) is

$$P = P_{ref} \times 10^{P(dB)/10} \tag{1.22}$$

As an example, if $P = 1000$ Watts, it can be expressed in decibel notation as

$$P(dBW) = 10 \log[P/(1 \text{ Watt})] = 30 \text{ dBW}$$

or

$$P(dBm) = 60 \; dBm$$

In converting equations like (1.17) to decibel notation, we note that quantities like cable losses, path loss, and antenna gains are already dimensionless power or power density ratios and may be converted directly to dB by taking their logarithm and multiplying by 10 without any further specification of a reference. Thus, Equation (1.17) in decibel notation becomes

$$L(dB) = P_t(dBW) + G_t(dBi) + G_r(dBi)$$

$$- L_t(dB) - L_r(dB) - P_s(dBW) \qquad (1.23)$$

[The notation *dBi* indicates that gain is evaluated with reference to an isotropic radiator.]

Thus, one can calculate the maximum allowable path loss by simple additions and subtractions. Such a calculation is called a link budget. Note that on the right-hand side of Equation (1.23), subtracting a quantity in dBW from another quantity in dBW produces a dimensionless result in dB.

As an example, consider a communications link in which the transmitter power is 100 Watts and the minimum acceptable received power is 10^{-12} Watts. Each of the cables attenuates the power by a factor of two. The transmitting antenna has a gain of 200, while the receiving antenna has a gain of 10. The corresponding link budget is as follows:

Transmitter power,	20 dBW
Transmitting antenna gain,	+23 dBi
Receiving antenna gain,	+10 dBi
Transmitter cable loss,	−3 dB
Receiver cable loss,	−3 dB
Minimum acceptable receiver power,	−(−120 dBW)

Maximum acceptable path loss, 167 dB

One then needs to develop an appropriate physical model relating path loss, L, to range, d, in order to determine the maximum range implied by $L_{max} = 167$ dB.

Finally we note that sometimes two signal strengths are compared by considering quantities that are proportional to the square root of power (e.g., electric field). In that case the equation for decibels is

$$E(dB) = 20 \log[E/E_{ref}]$$

This equation is equivalent to and will give exactly the same number of decibels as Equation (1.18).

1.3 Characteristics of Some Modern Communication Systems

1.3.1 Mobile Communications (Frequency Division Multiple Access, FDMA, and Trunking)

A primary concern in devising a mobile communication system is having a limited frequency bandwidth allocation while being required to provide access to a very large number of users each sending and receiving individual messages. The methods of managing the bandwidth allocation to best accomplish this task are broadly known as multiple access techniques. They include the following:

Frequency division multiple access (FDMA)
Trunking
Cellular telephone system layout for frequency reuse
Time division multiple access (TDMA)
Code division multiple access (CDMA)

Usually a number of multiple access techniques are used concurrently. TDMA and CDMA are used in systems employing digital signal modulation.

Although police vehicles began using mobile communication systems in the 1920s, the spread of such systems to the general consumer awaited the miniaturization of electronics that followed from the invention of the transistor after World War II. By the 1970s, the Bell Telephone Company in New York City had received a bandwidth allocation of 0.36 MHz for civilian mobile communications. This bandwidth was divided into twelve 30-KHz channels. Because different channels were used for the forward channels from the base station to the mobile and for the reverse channels from the mobile to the base station, the twelve 30-kHz channels provided only six duplex channels. Thus, *frequency division multiple access* (FDMA), the assignment of channels at distinct center frequencies to individual users, when used alone would only have allowed for six subscribers.

The customer base was increased by using *trunking*, the process of allocating channels to individual users from the pool of channels on a per call basis. Trunking calculations may be made using the Erlang-B equation:

$$p_B = \frac{U^N}{N! \sum_{m=0}^{N} [U^m/m!]} \tag{1.24}$$

where p_B is the probability of a call being blocked, U is the total caller traffic intensity in erlangs (1 erlang is equivalent to one subscriber making calls 100% of the time), and N is the number of duplex channels available for trunking,

A plot of this equation is shown in Figure 1.3. Applying this figure to the Bell Mobile phone system in New York City in the 1970s, if an acceptable blocking probability was chosen to be 0.1, for $N = 6$, this gives a traffic intensity 3.6 erlangs. It was determined that the typical subscriber in New York City in the 1970s required 0.0067 erlangs in the busy hour (i.e., the subscriber was

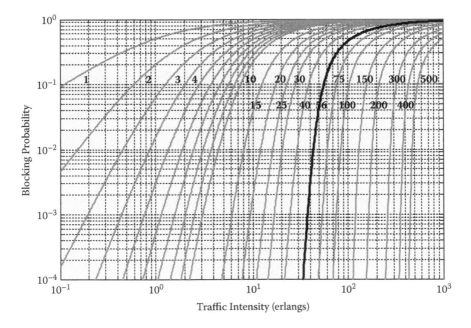

FIGURE 1.3 Erlang-B graph. Each curve is marked with the number of channels, N.

active 0.67% of the time). Thus, the number of subscribers that could use the six duplex channels was 3.6/0.0067 = 540 with a 10% probability of being blocked in the busy hour. If a smaller probability of being blocked is desired, it is clear that this will result in being able to handle a smaller number of subscribers. Also, the amount of busy hour traffic that each subscriber generates will vary with the particular service area and the year, and typically, it falls in the range 0.2% to 3% (i.e., 2 mE to 30 mE). In times of emergency (e.g., massive power failure, storms, terrorist attacks), the caller traffic may rise dramatically above the typical range resulting in a very high blocking probability and an effective breakdown of the mobile phone system. Values of the parameters in the Erlang-B equation have been tabulated over an extensive range and are displayed on the website http://www.quantumportal.com/erlangb.htm; the use of these tables is recommended for accurate Erlang-B calculations.

1.3.2 Analog Cell Phone Systems

In New York City in 1976, while only 540 subscribers could be accommodated, there was a waiting list of 3400 people who had applied for mobile telephones, and there was the opportunity to increase that number by orders of magnitude in an urban area with a population of over 10 million if a way could be found to accommodate them. It was unreasonable to expect that the problem could be solved by increasing bandwidth allocation alone. A new concept for accommodating a greatly expanded subscriber base was required.

To meet this need, in 1968, Bell Telephone Laboratories developed the *cellular telephone concept* that allows for accommodating an arbitrarily large number of subscribers with a fixed number of channels. The technology for implementing the cellular concept was available by the late 1970s. In 1983, the FCC allocated 660 duplex channels for the first U.S. cellular system, the Advanced Mobile Phone System (AMPS); the number of duplex channels was later expanded to 832. As depicted in Figure 1.4, the 832

824–849 MHz   869–894 MHz

Reverse Channels   Forward Channels
Mobile to Base Station   Base Station to Mobile

FIGURE 1.4 Assignment of frequency channels in AMPS.

reverse channels use the RF spectral range between 824 MHz and 849 MHz. The 832 forward channels use the spectral range between 869 and 894 MHz. For the sake of maintaining commercial competition, two companies were given frequency allocations in a given service area; Company B was the established land-line telephone provider, and Company A was a wireless service competitor. Each company received 416 duplex channels, of which 395 are used for telephone conversations while the remaining 21 are used for control data.

The cellular concept replaces a single high-power transmitter covering the entire service area (e.g., a particular city) with many low-power transmitters, each serving a small area (a cell) from a base station at the center of each cell. The decrease in required power was especially important for the mobile transmitter/receiver that could then be powered by small batteries and take the form of a lightweight handheld cell phone, which is in such widespread use today. Part of a matrix of hexagonal cells that would be superimposed on the map of the service area is depicted in Figure 1.5. Neighboring cells would be grouped into clusters; as an example, two clusters with seven cells in each cluster are shown in Figure 1.5. All the available channels would be assigned to each cluster, but each cell in the cluster would be allotted only a portion of the available channels; for a cluster of seven cells and with a total of 395 available duplex channels, each cell would be assigned 56 duplex channels. Adjoining cells would get different groups of channels to minimize interference. On the other hand, as shown in Figure 1.5, cells in different clusters with the same number designation would have the same group of channels assigned to them; such cells are called co-channel cells (e.g., the grey cells in Figure 1.5 are co-channel cells). Channel reuse is accomplished by repeating the clusters many times throughout

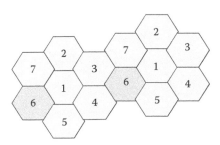

FIGURE 1.5 Two clusters with seven cells per cluster. Two co-channel cells are shown shaded.

the service area. If clusters are repeated M times in a service area and there are 395 duplex channels available, the effective number of channels in the service area is 395 M.

Each cell has a base station near its center with a tower-mounted antenna connected to transmitter and receiver units; these are in turn connected by cables to the Mobile Telephone Switching Office (MTSO), which the cell phone service provider maintains in the service area. The MTSO is in turn connected to the central telephone switching office.

The following sequence of events allows a cell phone to operate as part of a telephone network:

1. When a cell phone is turned on, it first listens for a system identification number to let it know either that service is available in its home system or that roaming charges will apply for using another system.
2. It then transmits a registration request that is passed on to the MTSO.
3. The MTSO keeps track of the mobile phone's location.
4. For an incoming call, the MTSO can then locate the phone in its database and ring it.
5. The MTSO will also select the duplex frequency channels to be used by the phone for either an incoming or outgoing call.
6. As shown in Figure 1.6, when the mobile moves to the edge of its cell, the decrease in signal strength at the base station whose cell it is leaving and the increase in signal strength at the base station whose cell it is entering, triggers a channel handoff (i.e., the MTSO assigns to the call a new duplex channel from those available in the new cell). This handoff is usually accomplished in such a seamless manner that the user is unaware of its occurrence.

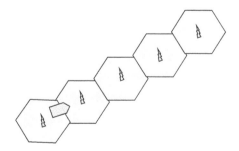

FIGURE 1.6 Handoff Geometry (not to scale). Cell radius ~ kilometers. Arrow indicates an automobile.

Antennas for both the base station and the mobile units should clearly be omnidirectional in the horizontal plane because the direction from the mobile to the base station cannot easily be determined and varies as the mobile moves. The antenna on the mobile is typically a small dipole antenna a quarter wavelength long with the antenna wire oriented vertically; a quarter of a wavelength at 840 MHz is 3 cm. The base station can accommodate an array of vertically oriented dipole antennas mounted on a tower; the array being used to achieve a larger gain. A base station in contact with a number of mobile phones is depicted in Figure 1.7.

One issue of concern in the cell phone frequency reuse scheme is that interfering signals from co-channel cells may be much weaker than in-cell signals; the larger distance traveled by the interfering signal is normally relied on to give a sufficient increase in path loss to reduce interference to an acceptable level. The ratio of distance traveled by a co-channel signal to distance traveled by an in-cell signal depends on the number of cells in a cluster, n_c, which can have only certain discrete vales for hexagonal cell geometry as given by

$$n_c = i_1^2 + i_1 i_2 + i_2^2 \tag{1.25}$$

where i_1 and i_2 are positive integers.

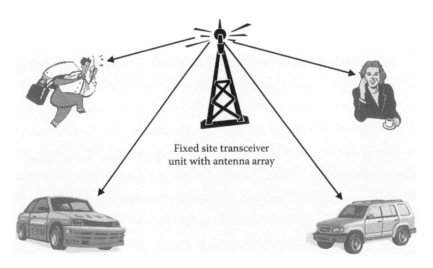

Fixed site transceiver
unit with antenna array

FIGURE 1.7 Base station antenna array, mounted on tower, communicates with mobile phones each with a small dipole antenna.

Thus, n_c may have the values 3, 4, 7, 12, 13, 19, 27, and so forth. The value of n_c determines, d_i, the minimum distance between co-channel cells divided by the cell average radius, r_c; namely,

$$d_i/r_c = (3n_c)^{1/2} \qquad (1.26)$$

It will be shown in Chapter 5 that for an urban service area, received signal power decreases as d^{-4} where d is the distance between the transmitter and the receiver. Thus, the ratio of in-cell signal power to interfering signal power is given by

$$P_c/P_i = (1/6)(d_i/r_c)^4 \qquad (1.27)$$

The divisor 6 indicates that there are six interfering cells with the same value of d_i as indicated in Figure 1.8. A technique known as sectorization can increase the ratio of in-cell power to interfering power. For example, if a cell were divided into six sectors each occupying only 60° of azimuthal angle, the factor of 1/6 in Equation (1.27) could be eliminated; base station antenna arrays can be designed to achieve this sectorization. The trade-off is 1/6 fewer channels available to be put into the trunking pool and a decrease in the number of subscribers who can be serviced.

By combining Equations (1.25) and (1.26), one can obtain the relationship between the minimum number of cells in each cluster

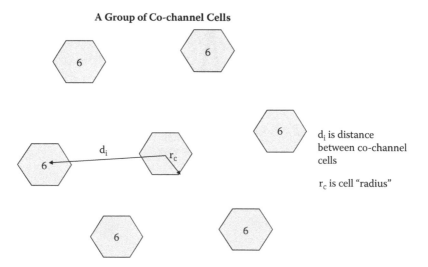

A Group of Co-channel Cells

d_i is distance between co-channel cells

r_c is cell "radius"

FIGURE 1.8 Six interfering cells.

and the minimum acceptable ratio of in-cell signal power to inter-fering power:

$$n_{c,min} = [(2/3)(P_C/P_i)]^{1/2} \tag{1.28}$$

For example, if the minimum acceptable value of P_C/P_i is 20 dB, then from Equations (1.28) and (1.25), $n_c = 12$. The number of channels in a cell available for trunking in AMPS is $N = 395/n_c$; for the present example $N = 395/12 = 33$.

1.3.3 Digital Cell Phone Systems (Time Division Multiple Access, TDMA, and Code Division Multiple Access, CDMA)

The AMPS cellular system was originally an analog system and used frequency modulation (FM). Hardware for the U.S. Digital Cellular (USDC) system was installed in 1991; allow-ance was made for the gradual replacement of analog channels with digital channels. By employing *time division multiple access* (TDMA) for the digital channels, a number of users (e.g., three or six) can be simultaneously supported in the same 30 kHz channel; with TDMA, the signals are compressed in time and each user employs cyclically repeating time slots that do not overlap with the time slots assigned to the other users. Digital modulation has other important advantages such as facilitat-ing the encryption of signals to prevent unauthorized message interception.

An alternative multiple access scheme used in digital systems is code division multiple access (CDMA) that replaces FDMA and TDMA. CDMA uses spread spectrum techniques so that many signals simultaneously occupy a large part of the available spectrum; however, the signals are coded so that they can be sepa-rated. Smaller signal power to interference power ratios can be tolerated in a CDMA. CDMA uses the available spectrum more efficiently and can accommodate a larger number of users with a given spectrum allocation.

As a simple example of CDMA, consider a system with two subscribers, each with a unique code assignment and simultane-ously receiving a different message. The messages are encoded using the codes of the intended receiver; then both messages are transmitted at the same time in the same undivided frequency channel. The combined transmission is then decoded at each receiver by applying that subscriber's code. The process is illus-trated in Table 1.3.

TABLE 1.3 A Simple Example of Code Division Multiple Access (CDMA)

	Subscriber #1	Subscriber #2
Code assigned	$C_1 = (1,-1)$	$C_2 = (1, 1)$
Message	$M_1 = 1,0,1,1$	$M_2 = 0,0,1,1$
Modified message $[0 \Rightarrow -1]$	$M_1{}^* = 1,-1,1,1$	$M_2{}^* = -1,-1, 1,1$
Encoded modified message	$C_1 \bullet M_1{}^* = (1,-1)\bullet(1,-1,1,1)$ $= (1,-1), (-1,1), (1,-1),$ $(1,-1)$	$C_2 \bullet M_2{}^* = (1,1)\bullet(-1,-1,1,1)$ $= (-1,-1), (-1,-1), (1,1), (1,1)$
Combined signal (interference signal)	$I = C_1 \bullet M_1{}^* + C_2 \bullet M_2{}^* =$ $(0,-2),(-2,0),(2,0),(2,0)$	
Decoded modified message	$C_1 \bullet I = (1,-1)\bullet[(0,-2),$ $(-2,0),(2,0),(2,0)] = 2,-2,2,2$	$C_2 \bullet I = (1,1)\bullet[(0,-2),(-2,0),$ $(2,0),(2,0)] = -2,-2,2,2$
Received message $[2 \Rightarrow 1]$ and $[-2 \Rightarrow 0]$	$M_1 = 1,0,1,1$	$M_2 = 0,0,1,1$

1.3.4 Overview of Past, Present, and Future Cell Phone Systems

The characteristics of some major mobile telephone systems in use during the years from 1979 to 2003 are listed in Table 1.4.

In Table 1.4, the three listings at the top are first-generation systems (1G) that used analog modulation. The next three listings are the second-generation (2G) systems employing digital modulation that enabled extensive signal processing procedures. The final three listings in the table are transitional between 2G systems and third-generation (3G) systems and are often referred to as 2.5G. These systems added high-speed data processing circuits such as enhanced data-rate for GSM evolution (EDGE) and have sufficient channel bandwidth to make data and video services available in addition to voice. The cell phone base stations were enabled to provide the enhanced services by direct connection to Internet access points.

The last entry in the table, global system for mobile (GSM), was accepted as a standard in Europe and most of Asia. In the United States, the use of different frequency bands and different multiple access techniques made some digital cellular systems incompatible

TABLE 1.4 Characteristics of Major Mobile Telephone Systems (1G to 2.5G)

System (Generation)	Usage Area	Year Introuced	Frequency Band (MHz)	Modulation	Channel Bandwidth (kHz)	Number of Channels	Multiple Access	Data Rate
NTT (1G)	Japan	1979	860–940	Analog/FM	25	600	FDMA	
AMPS (1G)	United States	1983	824–894	Analog/FM	30	832	FDMA	
E-TACS (1G)	Europe	1985	872–950	Analog/FM	25	1240	FDMA	
GSM (2G)	Europe	1990	890–960	Digital	200	124 8 users/channel	TDMA	
USDC (2G)	United States	1991	824–894	Digital	30	832 3 users/channel	TDMA	
PDC (2G)	Japan	1993	810–1513	Digital	25/12.5	1600 3/6 users/channel	TDMA	
IS95A (2G)	United States	1993	824–894	Digital	200	125 8 users/channel	CDMA	
PSC 1900 (2.5G)	United States/Canada	1995	1850–1990	Digital	200	299 8 users/channel	TDMA	
IS95B (2.5G)	United States	1999	824–894	Digital with MDR	1, 250	20 8 users/channel	CDMA	64 kbps
GSM (2.5G)	Europe/United States	2003	890–960	Digital with EDGE	200	124 8 users/channel	TDMA	384 kbps

Note: CDMA, code division multiple access; EDGE, enhanced data-rate for GSM evolution; FDMA, frequency division multiple access; FM, frequency modulation; GSM, global system for mobile; MDR, medium data rates; TDMA, time division multiple access.

with each other although multimode and multifrequency-band phones are becoming more common. The IS95B system employs CDMA, which uses the available range of the frequency spectrum more efficiently and is being widely adopted in 3G systems.

One important technological improvement introduced in 2.5G systems and carried over into 3G systems was packet switching in which transmitted data are gathered into suitably sized blocks (packets) that are transmitted with variable time delay as capacity becomes available. Because of the variable delay, packet switching is not suitable for real-time voice conversations, but it is acceptable for downloading data.

Third-generation (3G) cell phones, popularly known as smartphones, are characterized by a 500-fold increase in bandwidth compared with 2G to accommodate more advanced services including mobile e-mail messaging, texting, mobile Internet access, video calls, and mobile TV. A bandwidth of 15 to 20 MHz is required for 3G servers compared with only 30 to 200 kHz for 2G. The International Telecommunications Union (ITU) has issued specifications, with the designation International Mobile Telecommunications (IMT) 2000, for 3G mobile communications that include the following data transfer rates:

- 128 to 144 kilobits per second (kbps) for fast moving receivers (e.g., in a train or car)
- 384 kbps for receivers carried by pedestrians
- >2 megabits per second (Mbps) for fixed receivers

To accommodate 3G systems in Europe, Japan, and China, the spectral regions from 1900 MHz to 2025 MHz and 2110 MHz to 2200 MHz were reserved and auctioned off to telecommunication companies. The Universal Telecommunications System (UMTS) operating in these frequency bands became the successor to GSM. The first 3G system called Freedom of Mobile Multimedia Access (FOMA) became available in Japan on October 1, 2001, and was operated by the NTT DoCoMo company.

In North America and South Korea, the frequency bands used by UMTS were generally unavailable. In the United States, the FCC designated the following bands for 3G system use: 1710 to 1770 MHz, 2110 to 2170 MHz, and 2500 to 2690 MHz. The first 3G system in South Korea and the United States was CDMA-2000 1xEV-DO, a successor to the IS95B technology, which began operation in January 2002 in South Korea. Verizon

Wireless introduced a similar 3G system in the United States in October 2003.

In 2011, there are more than 5.3 billion mobile phone subscribers, a number equal to 70% of the world's population. Almost 20% of these subscribers (i.e., well over 1 billion) have fast mobile Internet access (3G or better). This enthusiastic acceptance of 3G ensured that further generations of cellular systems would be developed.

A new generation of cell phone systems seems to appear every 10th year since the 1G systems were first introduced around 1980. Systems advertising themselves as 4G began to appear as early as 2006 although they did not meet the 4G standard as specified by the ITU in 2008. Systems meeting the ITU 4G standard are expected to become available in 2011. The ITU data rates for 4G service are

- 100 Mbps for fast-moving receivers (e.g., in trains or cars)
- 1000 Mbps for slow-moving or fixed receivers

In most suggestions to date, for meeting the 4G goals, the CDMA spread spectrum technology is replaced by orthogonal frequency division multiple access (OFDMA). OFDMA is an advanced version of the CDMA spread spectrum concept, but users can be given different data rates by assignment of a different code spreading factor. In addition, 4G systems are expected to use multiple in/multiple out (MIMO) concepts that employ multiple antennas to achieve some measure of spatial division multiple access with additional multiplexing achieved through sensitivity to the direction in which signals are transmitted. Dynamic channel allocation and channel dependent scheduling are also expected to be features of 4G systems.

The voracious appetite for new wireless applications requires ever-increasing data transmission rates. This requirement when confronted by the limited spectral bandwidths that are available has led to impressive technological advances that enhance the efficiency with which spectral bandwidth is exploited. Nevertheless a fundamental limit is expressed in Shannon's equation. This puts a premium on controlling as much spectral bandwidth as possible.

In 2008, the U.S. government auctioned off portions of the UHF spectral region that are likely to be used for 4G systems. The auction offered five blocks of spectral bandwidth as follows:

Block A, 12 MHz bandwidth, 698 to 704 MHz and 728 to 734 MHz

Block B, 12 MHz bandwidth, 704 to 710 MHz and 734 to 740 MHz

Block C, 22 MHz bandwidth, 746 to 757 MHz and 776 to 787 MHz

Block D, 10 MHz bandwidth, 758 to 763 MHz and 788 to 793 MHz

Block E, 6 MHz (722 to 728 MHz)

The lions share including the coveted Block-C was purchased by Verizon for $9.4 billion. AT&T also bought a large swath of spectrum for $6.6 billion.

These huge sums effectively excluded competing by smaller cellular service companies.

Thus ironically, the breakup in the United States of the telephone system regulated monopoly, which at first generated many fiercely competing smaller companies, appears to be evolving into an unregulated duopoly of two communication giants. Both the regulated monopoly model and the model of unregulated competition between a multitude of smaller service providers led to many important innovations in communication science and technology and to effective constraint on service charges. It remains to be seen where an unregulated duopoly will lead.

1.3.5 Wireless Local Area Networks (WLANs) of Computers

In cell phone systems, the maximum distance between mobiles and a base station is in the range of 1 to 10 km. Much smaller ranges between transmitter and receiver apply to the case of wireless local area networks (LANs) of computers, usually 50 m or less indoors. Typically, networks are set up inside a building to link all the computers to a router and from there to a printer or to an access point for fast Internet connection. The use of wireless LANs is especially convenient for notebook computers that can access the Internet in various "hot zone" locations including many coffee bars. Each tabletop computer in a LAN will have a wireless PC card that contains an antenna, often a loop-type antenna. In wireless-enabled notebook computers, specialized antennas are built into the screen. "Wi Fi" LAN equipment operating at 2.4 GHz is the most popular, because a large variety of inexpensive components are available. Signal attenuation due to propagation through walls and floors is a dominating factor in designing a LAN. Both adequate signal strength within the local network and interference with neighboring co-channel networks need to be considered. Also, signal encryption is important to prevent unintended message interception.

1.3.6 SATCOM Systems

In contrast to the short transmission distances, in wireless LANs, intercontinental distances are involved in satellite communication systems (SATCOM). As mentioned in Section 1.1.2, long-distance communications based on reflection from the ionosphere are limited to lower frequencies (smaller than ~60 MHz). In modern long-distance communications the demands of a high rate of data transfer require large system bandwidth and therefore operation at high frequency in the microwave and millimeter wave spectral regions.

As will be derived in Chapter 8, the period of an artificial satellite in stable earth orbit depends on its altitude as given by the following relationship for a satellite with a circular orbital path:

$$T_e = 3.14 \times 10^{-7} R_a^{1.5} \tag{1.29}$$

where T_e is the period in seconds, and R_a is the distance of the satellite from the center of the earth.

From Equation (1.29), we calculate that satellites that revolve around the earth once every 24 hours have a corresponding distance from the center of the earth of 42,300 km. Such satellites are at an altitude of ~36,000 km above the earth's surface because the radius of the earth is 6375 km. These satellites are stationary with regard to an observer on the surface of the earth and are called geostationary orbit (GSO) satellites. GSO satellites are the type most frequently used in SATCOM systems. They have the obvious advantage that antennas aimed at them do not have to be reoriented. The satellites are always "in view," because their angular velocity matches that of the earth. Also, because of their large distance from the earth, as few as four satellites around the equator can cover most of the earth except for the polar regions.

SATCOM with GSO satellites is used for voice and data transmission to business locations with receiving antennas of at least 0.3 m in diameter. An example of such a GSO-SATCOM system is INTELSAT, which operates in two frequency bands centered at 1.6 GHz and 2.5 GHz. GSO-SATCOM is also widely used for transmitting television programming to home receivers with antennas about 0.45 m in diameter (direct broadcast satellite [DBS]). DBSs use digital modulation and circular polarization.

Another growing use of SATCOM is for aiding navigation of shipping and automobiles. A system of 24 global positioning satellites (GPSs) are in medium earth orbit (MEO) ~20,000 km above the earth and orbit the earth in about 12 hours. Signals from a GPS user at sea are received by three satellites, and tri-angulation is used to determine the position and speed of the transmitting vehicle. For GPS users on land, four satellites are used because altitude must also be determined. Atmospheric and ionospheric fluctuations can cause errors, but a differential GPS system can be used to improve accuracy. With differential GPS, a reference GPS receiver at a known location can provide error correcting information to other nearby GPS receivers, and position accuracy of 1 cm is possible. The received signal power from a GPS satellite is very low (~−130 dBm), and the receiving antenna has low gain (~0 dB) so that the signal power is typically smaller than the noise power, but digital signal processing techniques are used to extract the signal from the noise. GPS uses two frequency bands centered at 1.2276 GHz and 1.57542 GHz.

For smaller handheld antennas, mobile phone and other communication services can be provided to remote areas by a system of low earth orbit (LEO) satellites. An example of an LEO system is Iridium which was put into service in 1998 at a cost of ~$3.4 billion. Iridium employed 72 satellites at an elevation of 778 km operating at an RF frequency of 1.625 GHz. Such systems can have smaller path loss than GSO communication systems, but they are far more complex involving large numbers of satellites and introducing unwanted effects, which must be compensated for, such as a varying Doppler frequency shift due to the speed of the satellite with respect to the earth (~7 km/sec). The cost to subscribers for LEO satellite service was very high (~$2/minute) and the equipment was expensive to acquire (~$6000 for a satellite phone); as a result the initial subscriber base was smaller than expected, and Iridium declared bankruptcy in August 1999. However, the system was acquired by another operator at nominal cost and is now being successfully operated at a profit. The unique communications opportunity such a system provides to such people as newsmen and sportsmen in otherwise inaccessible locations often justifies the relatively high cost of subscribing.

Finally, we note that there is a growing trend to develop SATCOM systems at millimeter wavelengths. The wider bandwidth available here for faster rate of data transmission is clear. However, special effects such as strong signal attenuation by rain must be taken into account. The U.S. Department of Defense operates its GSO military strategic and tactical relay (MILSTAR) systems in the EHF band (30 to 300 GHz). Because of the difficulty of disposing of waste energy in space, a premium is placed on high efficiency of the transmitter RF power source in space; wide bandwidth vacuum traveling wave tubes (TWTs) with depressed collectors are inherently more efficient than solid-state sources, and the TWTs have become the preferred space-borne transmitter power amplifier especially at millimeter wavelengths.

1.4 The Plan of This Book

Chapter 2 treats the subject of noise, a ubiquitous physical phenomenon, that ultimately limits the performance of any communication system. Starting from a basic consideration of the sources of noise, useful approximations for estimating the noise power in practical systems such as cell phones and SATCOM will be derived. To evaluate the received signal power that must ultimately overcome the noise, one needs to analyze both antenna structures and the propagation losses along the path between the transmitting and receiving antennas. Chapter 3 deals with understanding important classes of antennas, while Chapter 4 deals with antenna arrays. Propagation in free space, propagation above a planar earth, and propagation in an urban environment are all considered in Chapter 5. Chapter 6 shows how the understanding of antennas and propagation can be applied to the design of cell phone systems and wireless LANs taking statistical variations into account. Chapter 7 describes schemes for long-range communications that exploit the properties of the troposphere and the ionosphere. Finally, Chapter 8 considers the design of SATCOM systems of both the GSO and LEO varieties.

There are also three useful appendices. Appendix A provides a Table of Physical Constants. Appendix B defines the operators in both Cartesian and spherical coordinate systems. Appendix C gives the expressions for differential line, area, and volume elements in Cartesian and spherical coordinates. Students will find the appendices helpful especially when solving the problem sets that appear at the end of each chapter.

Problems

1.1. The magnetic field of a plane wave propagating in free space is given by the following expression:

$$H(r,t) = 0.2 \cos(6.28 \times 10^9 [t - x/c] + \pi/4) \, a_z \, (\text{amps/m})$$

    a. Write a similar expression for the electric field.
    b. What are the frequency and the wavelength?
    c. What is the direction of propagation and the direction of linear polarization?

1.2. Calculate the capacity (in bps) of communication channels with the following parameters:

    a. Center frequency, $f_0 = 85$ kHz; Bandwidth, $B = 0.01 \, f_0$; S/N = 10
    b. Center frequency, $f_0 = 850$ MHz; Bandwidth, $B = 0.01 \, f_0$; S/N = 10
    c. Center frequency, $f_0 = 850$ MHz; Bandwidth, $B = 0.01 \, f_0$; S/N = 2

1.3. A base station antenna with a gain of 16 dBi is supplied with 5 Watts. Calculate the EIRP.

1.4. Effective noise input power into a mobile receiver is –100 dBm and a signal-to-noise ratio of at least 5 dB is required for acceptable quality of reception. Calculate the maximum acceptable propagation loss, given that the transmitter power is 10 Watts, the transmitter feeder loss is 6 dB, the base station antenna gain is 12 dBi, the mobile antenna gain is 1 dBi, and the mobile feeder loss is 1.5 dB.

1.5. RF power of 100 Watts is radiated isotropically in free space. Calculate the power density 3 km from the source. Making a local plane wave approximation, calculate the values of electric field and magnetic field that this corresponds to.

1.6. What are the power density and field strengths that might be realized in Problem 1.5 if a transmitting antenna with a gain of 15 dBi were used?

1.7. How many cells are required in a cell phone system cluster if the minimum acceptable ratio of in-cell power to interfering power is 20 dB?

1.8. A cell has 2000 users who generate an average busy-hour traffic of 2.5 mE each. How many channels are needed to serve these users with blocking probability no greater than 4%?

1.9. With the same number of channels per cell as calculated in Problem 1.8, if some emergency were to cause the average traffic to rise to 30 mE per subscriber, what would be the new value of probability of a blocked call? (Comment on the expected performance of a cell phone system during an emergency when an unusually large number of subscribers are trying to call home.)

1.10. With the same number of channels per cell as calculated in Problem 1.8 and with the same blocking probability of 4%, how many subscribers can be accommodated if the cell is divided into three sectors?

1.11. Perform a first-order layout design of a cell phone system to cover a metropolitan service area with 1.5 million subscribers and a total area of 64 sq km (i.e., calculate the number of cells in a cluster, the number of clusters needed to cover the service area, and the radius of each cell). Choose the following service quality parameters:

Minimum signal-to-interference power ratio = 25 dB
Maximum blocking probability = 0.3%
Number of "busy hour" erlangs per subscriber = 0.006

1.12. Construct a table similar to Table 1.3 showing how CDMA may be used to transmit and separate the following messages: $M_1 = 0,1,0,0$ and $M_2 = 1,1,0,1$.

Bibliography

1. N. Gehani, *Bell Labs: Life in the Crown Jewel* (Silicon Press, Summit, NJ, 2003).

2. S.R. Saunders, *Antennas and Propagation for Wireless Communication Systems* (John Wiley & Sons, Chichester, UK, 1999), 1–20.

3. T.S. Rappaport, *Wireless Communications: Principles and Practice.* 2nd ed. (Prentice Hall PTR, Upper Saddle River, NJ, 2002).

4. D.M. Pozar, *Microwave and RF Wireless Systems* (John Wiley & Sons, New York, 2001).

5. A.W. Hirshfield, *The Electric Life of Michael Faraday* (Walker and Company, New York, 2006).

6. D.P. Agrawal and Q.-A. Zeng, *Introduction to Wireless and Mobile Systems.* 3rd ed. (Cengage Learning, Stamford, CT, 2011).

7. C.D. Trowbridge Jr., *Marconi: Father of Wireless, Grandfather of Radio, Great-Grandfather of the Cell Phone. The Story of the Race to Control Long-Distance Wireless* (BookSurge, Charleston, SC, 2010).

CHAPTER **2**

Noise in Wireless Communications

We saw in Chapter 1 that the information capacity of a communications channel depends on the ratio of signal power to equivalent noise power at the receiver input. The present chapter will describe the sources of noise power and develop expressions for calculating the noise power. It will be seen that the calculations depend on such factors as the central frequency and bandwidth of the receiver, on noise generation by objects in the region that the antenna is focused onto, and on noise generation in the receiver itself.

2.1 Fundamental Noise Concepts

2.1.1 Radiation Resistance and Antenna Efficiency

Antenna Ohmic efficiency, η, was mentioned in Section 1.2.3. To elaborate, we introduce the concept of radiation resistance, R_r. When a net amount of power is coupled from a transmitter into an antenna, it is either dissipated in Ohmic losses in the imperfect conductors comprising the antenna, or it is radiated. As shown in Figure 2.1, we consider a dipole antenna fed by a current:

$$I(t) = I_o \cos(\omega t + \alpha) \tag{2.1}$$

The time-averaged power dissipated by the Ohmic resistance of the antenna, R_o, is given by

$$P_o = \tfrac{1}{2} I_o^2 R_o \tag{2.2}$$

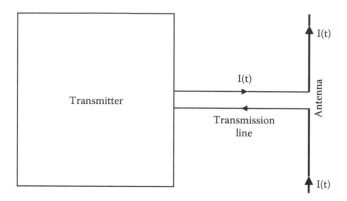

FIGURE 2.1 Dipole transmitting antenna.

Analogously, the time-averaged power radiated by the antenna is given by

$$P_r = \tfrac{1}{2}I_o^2 R_r \qquad (2.3)$$

Thus, we see that R_r represents a resistance that would dissipate the same amount of power as the antenna radiates when the current in that resistance equals the antenna input current. The antenna equivalent circuit is then two resistors in series, R_o and R_r. There are also reactive elements in the equivalent circuit, but these are usually compensated for by adding external inductance or capacitance if necessary, so that at the operating frequency the capacitive and inductive reactive effects cancel. The equivalent circuit for a compensated dipole transmitting antenna is simply the resistors R_o and R_r in series as depicted in Figure 2.2.

Antenna efficiency accounting for Ohmic losses, designated by η, is the ratio of radiated power to the total power fed into the antenna. It is given by

$$\eta = P_r/(P_o + P_r) = R_r/(R_o + R_r) \qquad (2.4)$$

For many large antennas Ohmic efficiency is close to 100%, but this may not be true if the antenna dimensions are small compared with the wavelength of the radio frequency (RF) signal.

2.1.2 Nyquist Noise Theorem, Antenna Temperature, and Receiver Noise

In 1928, Harry Nyquist (1889 to 1976, b. Nilsby, Sweden), working at Bell Telephone Laboratories, derived the expression for

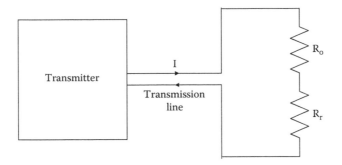

FIGURE 2.2 Transmitting antenna equivalent circuit.

noise voltage in a frequency bandwidth, Δf arising from the thermal motion of electrons in a resistor of resistance R at absolute temperature T:

$$V_{N,rms} = (4k_B T \, \Delta f \, R)^{½} \qquad (2.5a)$$

where $k_B = 1.38 \times 10^{-23}$ Joules/°K is the Boltzmann constant.

Equation (2.5a) is equivalent to stating that the noise power that would be delivered by a resistor R at temperature T to a receiver with matched input resistance ($R_i = R$) is

$$P_N = k_B \, T \, \Delta f \qquad (2.5b)$$

To appreciate the equivalence of Equations (2.5a) and (2.5b), consider the equivalent circuit in Figure 2.3 where a resistance R at temperature T has been replaced by a noiseless resistance at $T = 0$°K in series with a noise generator having voltage V_N. The resistance and noise generator are connected across the input terminals of an amplifier with input resistance, $R_i = R$. The voltage appearing across R_i is $V_i = V_N/2$. The average noise power at the input of the amplifier is

$$P_N = <V_i^2>/R_i = <V_N^2>/(4 \, R)$$

or

$$<V_N^2> = 4 \, R \, P_N = 4 \, R \, k_B \, T \, \Delta f \qquad (2.5c)$$

where we have used Equation (2.5b).

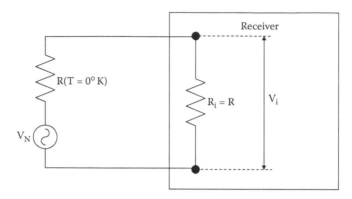

FIGURE 2.3 Gedanken experiment for defining Nyquist noise voltage of a resistor, R, at temperature, T.

Taking the square root of both sides of Equation (2.5c) clearly yields the expression for the Nyquist noise voltage in Equation (2.5a).

The noise described by Equation (2.5) is called Nyquist noise, or Johnson noise, or sometimes simply thermal noise. It is white noise (i.e., the noise power is linearly proportional to the bandwidth but is independent of the central frequency). Equation (2.5) is valid only in the classical limit where quantum mechanical effects are unimportant, which holds if

$$\hbar\, \omega \ll k_B\, T \qquad (2.6)$$

where $\hbar = 1.05443 \times 10^{-34}$ Joule seconds is Planck's constant divided by 2π.

For ambient temperature ($T_0 = 290°K$), and frequency in the microwave or millimeter-wave bands, inequality (2.6) is satisfied (i.e., we are in the classical limit and the thermal noise is white).

In addition to the Nyquist noise associated with the Ohmic resistance of an antenna, R_0, there will be noise picked up by the antenna from external sources. This is usually accounted for by assigning an antenna temperature, T_A, to the antenna radiation resistance, R_r. The noise power picked up by an antenna (and therefore the antenna temperature) will vary with the RF frequency, the temperature of the objects in the viewing field of the antenna, attenuation of the noise by atmospheric absorption, and so forth. The various contributions to antenna temperature will be discussed in some detail in Section 2.2.

In addition to the Ohmic noise and pickup noise contributions, a third source of noise is the receiver amplifier. The amplifier is characterized by its power gain, G; its bandwidth, Δf; and its noise factor, F. Noise factor is usually defined with reference to a gedanken experiment in which a white noise source at the ambient temperature, T_o, is applied to the amplifier input and produces a noise power at the output of the amplifier that is equal to

$$P_{N,\,out} = G\ F\ P_{N,in} = G\ F\ T_o\ k_B\ \Delta f \qquad (2.7)$$

For a noiseless amplifier $F = 1$, but for a real amplifier F will be larger than 1. The noise figure of the amplifier $F(dB) = 10\ \log F$. The reader should note that Equation (2.7) is only valid when $P_{N,in} = T_o\ k_B\ \Delta f$ and not for an arbitrary noise input.

To treat an arbitrary value of input noise together with the noise generated by the amplifier, one represents the amplifier by the equivalent model of a noiseless amplifier (i.e., an amplifier with $F = 1$) with an extra noise source at the input at a temperature T_e. Then, for the case $P_{N,in} = T_o\ k_B\ \Delta f$,

$$P_{N,\,out} = G\ (P_{N,in} + T_e\ k_B\ \Delta f) = G\ (k_B\ T_o\ \Delta f + k_B\ T_e\ \Delta f) \quad (2.8)$$

Comparing Equations (2.7) and (2.8) gives the following relationship between T_e and F:

$$T_e = (F - 1)\ T_o \qquad (2.9)$$

A realistic noisy amplifier may be represented by an ideal noise-free amplifier with input noise generated by the amplifier input resistance, R_i, at a temperature $(F - 1)\ T_o$.

Next, consider noise for the case of two amplifiers connected in series (i.e., cascaded) as depicted in Figure 2.4. The first amplifier has gain G_1 and noise factor F_1, while the second amplifier has gain G_2 and noise factor F_2. The bandwidth of the amplifiers is assumed to be the same and is denoted by Δf. A white noise source at the ambient temperature is applied to the input of amplifier number 1. Then, referring to Equation (2.7), we see that the output of amplifier 1 is

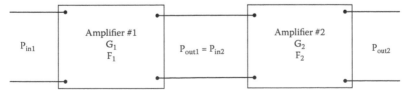

FIGURE 2.4 Two cascaded amplifiers.

$$P_{out,1} = G_1 \, F_1 \, T_o \, k_B \, \Delta f = P_{in,2} \tag{2.10}$$

As indicated in Equation (2.10) the output of amplifier 1 becomes the input to amplifier 2. Then, referring to Equations (2.8) and (2.9), we see that the output of amplifier 2 is

$$P_{out,2} = G_2 \, (P_{in,2} + T_{e2} \, k_B \, \Delta f)$$

$$= G_2 \, (G_1 \, F_1 \, T_o \, k_B \, \Delta f + (F_2 - 1) \, T_o \, k_B \, \Delta f)$$

$$= G_1 \, G_2 \, [F_1 + \frac{F_2 - 1}{G_1}] \, T_o \, k_B \, \Delta f$$

Thus, for the two-amplifier cascade, the overall gain is $G_1 \, G_2$ and the overall noise factor is

$$F_1 + \frac{F_2 - 1}{G_1}$$

For n cascaded amplifiers having gains G_1, G_2, G_3, ..., G_n and noise factors F_1, F_2, F_3, ..., F_n, the overall gain is

$$G = G_1 \times G_2 \times G_3 \times \ldots \times G_n \tag{2.11a}$$

while the overall noise factor is

$$F = F_1 + \frac{F_2 - 1}{G_1} + \frac{F_3 - 1}{G_1 G_2} + \ldots + \frac{F_n - 1}{G_1 G_2 \, x - x \, G_{n-1}} \tag{2.11b}$$

We see from Equation (2.11b) that because all the gains are likely to be much larger than unity, the noise contribution from the first stage will be dominant.

2.1.3 Equivalent Circuit of Antenna and Receiver for Calculating Noise

In line with the preceding discussion, equivalent circuits of a receiving antenna and receiver amplifier for the purpose of calculating the input noise power are shown in Figures 2.5 and 2.6. In Figure 2.5, the components representing antenna reactance, inductance L and capacitance C, are shown explicitly, but these are usually compensated for by adding reactive components external to the antenna so that the reactive impedance at the operating

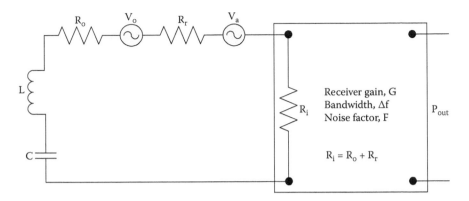

FIGURE 2.5 Equivalent circuit of an antenna feeding a noisy amplifier.

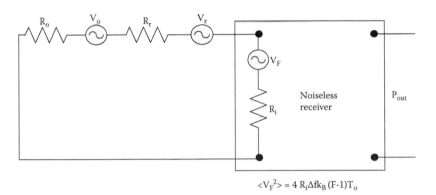

$$\langle V_F^2 \rangle = 4\, R_i \Delta f k_B\, (F-1) T_o$$

FIGURE 2.6 Equivalent circuit of an antenna feeding an amplifier. The noisy amplifier has been replaced by a noiseless amplifier plus an additional noise source at the input. The antenna reactance is assumed to be compensated in the operating frequency band.

frequency is zero; thus, L and C have been omitted in Figure 2.6. Also in Figure 2.5, a noisy receiver is shown with its noise figure, F. In Figure 2.6, the receiver is noiseless, and the effect of amplifier noise has been represented by the voltage V_F applied at the amplifier input.

V_o is the Nyquist voltage across the Ohmic resistance of the antenna, R_o, at the ambient temperature T_o; that is,

$$\langle V_o^2 \rangle = 4\, \Delta f\, R_o\, k_B\, T_o \qquad (2.12)$$

V_A is the antenna noise voltage due to noise from external sources picked up by the antenna, and we may write

$$\langle V_A^2 \rangle = 4\, \Delta f\, R_r\, k_B\, T_A \qquad (2.13)$$

where T_A is called the antenna temperature.

The third noise voltage in Figure 2.6, V_F, is due to noise generated by the receiver amplifier itself and in line with the preceding discussion:

$$\langle V_F^2 \rangle = 4 \, \Delta f \, R_i \, k_B \, (F - 1) \, T_o \tag{2.14}$$

The equivalent noise voltage across the resistor R_i at the amplifier input due to all three noise generators in Figure 2.1 is $V_i = \frac{1}{2}(V_o + V_A + V_F)$, or the time-averaged noise power delivered to the amplifier input is

$$P_N = \langle V_i^2 \rangle / R_i = \frac{1}{4} \langle (V_o + V_A + V_F)^2 \rangle / R_i$$

and because the three noise voltages are uncorrelated:

$$P_N = (\frac{1}{4} \langle V_o^2 \rangle + \frac{1}{4} \langle V_A^2 \rangle + \frac{1}{4} \langle V_F^2 \rangle) / R_i$$

Then, using Equations (2.12), (2.13), and (2.14) that relate the mean square noise voltages to the temperatures, we get

$$P_N = k_B \, \Delta f \, [R_o \, T_o + R_r \, T_A + R_i \, (F - 1) \, T_o] / R_i \tag{2.15}$$

Now, assuming impedance matching (i.e., $R_i = R_o + R_r$), and recalling that radiation efficiency of the antenna, $\eta = R_r / (R_o + R_r)$, Equation (2.15) may be written as

$$P_N = k_B \, \Delta f \, [(1 - \eta) \, T_o + \eta \, T_A + (F - 1) \, T_o] \tag{2.16}$$

The first term inside the square brackets in Equation (2.16) is due to Ohmic resistance of the antenna, the second term is due to noise pickup by the antenna from external sources, and the third term is due to noise generation by the receiver amplifier. We will use Equation (2.16) to calculate equivalent noise input power, but first we must be able to evaluate the antenna temperature, T_A.

2.2 Contributions to Antenna Temperature

Contributions to antenna temperature are made both by thermal sources and nonthermal sources such as sunspots or molecular resonances in atmospheric gases. The nonthermal sources produce noise that usually varies more strongly across the frequency spectrum than thermal noise. The contributions are often put into four categories as indicated by the following equation:

$$T_A = T_s + T_b + T_{cosmic} + T_{atmospheric} \qquad (2.17)$$

The first term in Equation (2.17) is the temperature of objects, which lie in the field of view of the antenna. For systems such as cell phones or pagers with antennas focused along the horizontal plane, to good approximation, $T_s = T_0$ the ambient temperature of buildings, trees, cars, people, and so forth.

The second term in Equation (2.17) is due to the "big bang" noise, a residual microwave noise remaining from the calamitous beginning of the universe. Because T_b is relatively small it usually does not have to be accounted for in designing most communication systems. However, discovery of the big bang noise was made by communication system scientists, and the interesting story of their experiments will be described in Section 2.2.4.

The last two terms in Equation (2.17) represent, respectively, noise from cosmic sources and from atmospheric gases. These are nonthermal sources, and their noise spectra are strongly frequency dependent.

2.2.1 Thermal Sources of Noise and Blackbody Radiation

Max Planck in 1902 successfully derived the expression for the frequency spectrum of noise emitted from a blackbody at a temperature *T*. A blackbody is an idealized object that absorbs all incident electromagnetic radiation, but Planck's law, Equation (2.18), is approximately valid for any opaque physical object. Planck's law may be written as

$$I(f, T) = 2 h f^3 / [c^2 \{exp (hf/k_B T) - 1\}] \qquad (2.18)$$

where $I(f,T)$ is radiation intensity (i.e., the power radiated per unit frequency interval, per unit area, per unit time), and h is Planck's constant (h = 6.62518×10^{-34} Joule-seconds).

In deriving his law, Planck for the first time introduced the concept that light energy was exchanged with atoms only in multiples of a discreet energy, $W = hf$. The energy packet, *hf*, was called a photon, and Planck's law initiated the development of quantum mechanics. Quantum mechanics is the body of scientific principles that attempts to explain the behavior of matter and its interactions with energy on the scale of atomic particles.

In Figure 2.7, Planck's law is plotted for the temperatures $T = 3000°K$, $4000°K$, and $5000°K$. Each curve has a maximum at a frequency given by

$$f_{max} = 59 \, T \quad GHz \tag{2.19}$$

To the right of the peaks the curves flatten.

The limit of Planck's law when $hf \ll k_B T$ is relevant to most situations in wireless communications. If we apply this limit to Equation (2.18) and accordingly use the approximation

$$\exp(hf/k_B T) \approx 1 + (hf/k_B T)$$

Equation (2.17) becomes

$$I(f,T) \approx 2k_B T/\lambda^2 \tag{2.20}$$

Note that dependence on Planck's constant has vanished indicating that Equation (2.20) is a classical limit in which quantum effects are not observed.

To derive an expression for the noise power received by an antenna and its associated receiver using Equation (2.20), $I(f,T)$ must be multiplied by the bandwidth of the receiver, Δf, by the solid angle and subtended by the receiving antenna, Ω, and by the

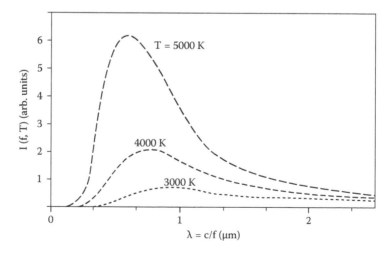

FIGURE 2.7 Spectra of blackbody radiation for temperatures of 3000 K, 4000 K, 5000 K.

area of the blackbody which lies within the beam of the receiving antenna, A_b. That is,

$$P_N = 2\, k_B\, T\, \Delta f\, \Omega\, A_b/\lambda^2 \qquad (2.21)$$

Referring to Figure 2.8, we set the solid angle as equal to

$$\Omega = A_e/R^2 \qquad (2.22)$$

where A_e is the effective area of the antenna, and R is the distance from the antenna to the blackbody.

Also, the area of the blackbody which fills the antenna beam is

$$A_b = BW_\phi\, BW_\theta\, R^2 \approx 4\pi\, R^2/D \qquad (2.23)$$

where we have used Equation (1.9).

Then, assuming that antenna Ohmic losses are negligible, we can use Equation (1.15) to further reduce Equation (2.23) to

$$A_b \approx R^2\lambda^2/A_e \qquad (2.24)$$

Finally, after substituting Equations (2.22) and (2.24) into Equation (2.21), we get

$$P_N \approx 2\, k_B\, T\, \Delta f \qquad (2.25)$$

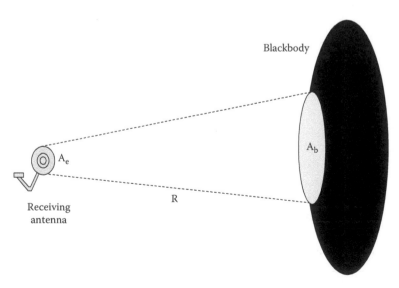

FIGURE 2.8 Configuration showing thermal noise pickup by an antenna.

The noise power from the blackbody will be equipartitioned between two orthogonal polarizations (i.e., directions of the electric field) that are normal to the direction of propagation. Thus, for an antenna that accepts only one polarization the noise power received is

$$P_N \approx k_B T \, \Delta f \tag{2.26}$$

Equation (2.26) is of the same form as the equations we have been using for thermal noise, for example, Equation (2.5b). This implies the following: that the thermal noise is due to electromagnetic noise radiation from an opaque object at temperature T which fills the field of view of the receiving antenna; that any frequency received is in the classical limit ($hf \ll k_B T$), and that the antenna is polarized.

2.2.2 Cosmic Noise

In Figure 2.9, the values of antenna temperature needed to account for noise from galactic sources and from atmospheric gases are plotted as a function of frequency.[4] The galactic noise arises from celestial sources in the Milky Way galaxy, outside the earth's solar system, such as radio stars and nebulae. Galactic noise is one component of cosmic noise. It is dominant at frequencies below 300 MHz. Useful approximations for estimating antenna temperature range due to galactic noise are

$$T_{galactic,\ max} = 6.2 \times 10^{21}\ f^{-2.3} = 12.4\ \underline{f}^{-2.3} \tag{2.27a}$$

and

$$T_{galactic,\ min} = 5.0 \times 10^{20}\ f^{-2.3} = \underline{f}^{-2.3} \tag{2.27b}$$

where temperature is in degrees Kelvin, $\underline{f}$ is the frequency in Hz, and $\underline{f}$ is a dimensionless frequency normalized to 1 GHz.

Of course, the galactic noise temperature is variable and at any given time may be expected to fall somewhere between the limits given by Equations (2.27a) and (2.27b).

The maximum galactic noise temperature is sometimes stated in a very approximate but more easily remembered form as

---

[4] D. C. Hogg and W. W. Mumford, "The Effective Noise Temperature of the Sky," *Microwave Journal 3*, March 1960, p.80

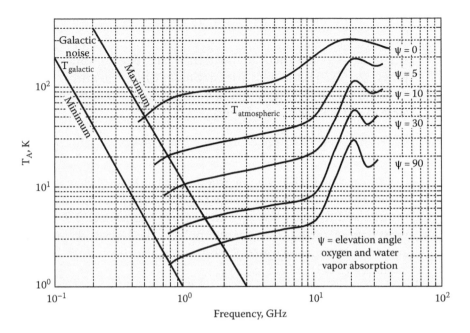

FIGURE 2.9 Contributions to antenna temperature from atmospheric gases and from galactic noise.

$$T_{galactic,\,max} \cong T_o \underline{\lambda}^{2.3} \qquad (2.27c)$$

where $\underline{\lambda}$ is a dimensionless wavelength normalized to 1 m.

In addition to the galactic noise, the sun or the moon may occasionally be in the field of view of a receiver antenna that is directed skyward; they would then contribute additional cosmic noise. The sun's noise temperature varies with such factors as sunspot activity, but on the average may be approximated by

$$T_{sun} = T_o \times 675/\underline{f} \qquad (2.28a)$$

The moon's noise temperature is approximately

$$T_{moon} = 170°K \qquad (2.28b)$$

Both the galactic noise and noise contributions from the sun or the moon must travel through the earth's atmosphere before reaching the receiving antenna and so are subject to attenuation by atmospheric gases and rain. The total value of this attenuation in dB is

$$L(A) = L(G) + L(R) \qquad (2.29)$$

where $L(G)$ is the total attenuation in dB due to atmospheric gases, and $L(R)$ is the total attenuation in dB due to rain.

Then, the total contribution to antenna noise temperature from sources outside the atmosphere is

$$T_{cosmic} = (T_{galactic} + u\, T_{sun/moon})\, 10^{-L(A)/10} \qquad (2.30)$$

In Equation (2.30), u is the fraction of the field of view area of the receiving antenna that is filled by the sun or the moon. Both the sun and the moon occupy an angular diameter of $0.5°$ when viewed from earth. Thus, for example, if the beamwidth of the antenna pattern encompassing the sun or the moon were $1°$, the factor $u = (0.5/1)^2 = 0.25$.

$L(G)$ for elevation angles of $\psi = 90°, 30°, 10°, 5°$, and $0°$ are plotted as a function of frequency in Figure 2.7. At other elevation angles $y > 5°$,

$$L(G) = L(G)_{zenith}/\sin\psi \qquad (2.31)$$

where $L(G)_{zenith}$ is the value of $L(G)$ when $\psi = 90°$.

For frequency <3 GHz, $L(R)$ in Equation (2.29) may be considered negligibly small. For higher frequencies, such as are found in many SATCOM systems, $L(R)$ cannot be neglected, and it will be considered in detail in Chapter 8.

2.2.3 Atmospheric Noise

The family of curves in Figure 2.9 that rise with frequency may be used to evaluate the contribution of atmospheric gases to antenna temperature. The peak near 22 GHz is due to a molecular resonance in water vapor, H_2O. Other peaks occur at higher frequencies due to resonances in both water vapor and molecular oxygen, O_2. The variation in the atmospheric contribution to antenna temperature with elevation angle is due to the fact that the atmosphere is "optically thin," and signal attenuation by atmospheric gases varies with path length. The contribution of atmospheric gases to antenna temperature when $L(G) \gg L(R)$ is

$$T_{atmospheric} = T_{gases}(1 - 10^{-L(G)/10}) \qquad (2.32)$$

This quantity is represented by the family of rising curves in Figure 2.7. Each value of elevation angle ψ corresponds to a different value of path length and hence a different value of attenuation by the gases, $L(G)$. When elevation angle y = 0° in the horizontal plane, the path length and the atmospheric noise temperature are maximum; this situation applies to cell phones.

For cell phones, attenuation data in more useful form may be found in Figure 2.10. In Figure 2.10, specific attenuation (dB/km) by atmospheric gases and rain is shown for a horizontal path.

Note that there is a range of frequency between ~500 MHz and ~2 GHz where both cosmic noise temperature and atmospheric noise temperature at y = 0° are both below 100°K. This frequency range has been chosen for cell phone system operation partly because of the low noise.

SATCOM systems are usually designed to continue operation in the presence of heavy rain. In that case, $L(R) \gg L(G)$, and in place of Equation (2.32), one calculates the antenna noise contribution arising in the atmosphere from

$$T_{atmospheric} = T_{rain}(1 - 10^{-L(R)/10}) \qquad (2.33)$$

where the noise temperature of rain is T_{rain} = 260°K.

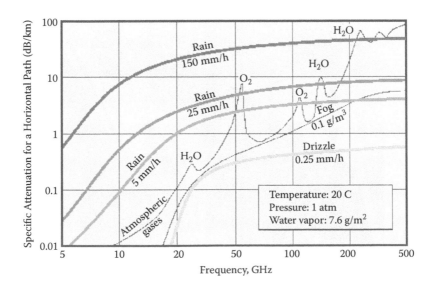

FIGURE 2.10 Specific attenuation (dB/km) of microwaves by atmospheric gases and rain.

2.2.4 Big Bang Noise (Cosmic Microwave Background Radiation)

In 1964, Arno Penzias and Robert Wilson, two young radio astronomers, were attempting to study the faint radio emissions from gas in the Milky Way galaxy using a very large horn antenna at the Crawford Hill in Holmdel, New Jersey, a location of Bell Telephone Laboratories (BTL). A photograph of Wilson (on the left) and Penzias together with the antenna they used may be seen in Figure 2.11; this antenna was 20 feet wide and had been previously used by other BTL researchers to detect signals from Echo, the first communications satellite. Penzias and Wilson were using as their receiver amplifier a traveling wave maser tuned to amplify a very narrowband signal centered at 4.08 GHz. At this frequency, they measured at zenith elevation ($\psi = 90°$) atmospheric noise radiation of 2.3°K in rough agreement with the measurements of Hogg and Mumford presented in Figure 2.9; atmospheric noise could be clearly identified because of its variation with the elevation angle of the antenna. (The Hogg and Mumford data should be regarded as average values; more specific values of atmospheric noise will depend on local conditions such as humidity.) Penzias and Wilson also calculated that the Ohmic losses in their antenna contributed an additional 0.8°K to the noise temperature—that is, in Equation (2.16), the term, $(1 - h) T_o = 0.8°K$. Liquid helium cooling was used in the receiver to keep receiver noise at a negligible level—that is, in Equation (2.16), the term $(F - 1)T_o$ was negligible. The radio astronomy studies were thus expected to be

FIGURE 2.11 Arno Penzias and Robert Wilson standing beside the antenna that they used in discovering the universe's microwave background radiation.

limited by noise temperature of only ~3°K. The measurements made by Penzias and Wilson, however, corresponded to a total noise temperature of 6.7°K and an antenna temperature of 5.8°K ± 1°K; the unexpected excess noise appeared to come from background radiation arriving with equal intensity from all directions and corresponding to a noise temperature of 3.5°K ± 1°K.

At first, the BTL researchers were frustrated and puzzled by this unexpected noise that was obscuring their radio astronomy studies. They tried many drastic measures to eliminate the extra noise including covering all the seams in the antenna with aluminum tape and cleaning pigeon droppings on the antenna surface; all to no avail.

The explanation for the background microwave radiation came when Penzias learned that Professor Robert Dicke and his research group at Princeton University, New Jersey, were preparing an experiment to measure cosmic microwave background radiation as a test of cosmological theories for the creation of the universe. Penzias called Dicke and described the experimental measurements that he and Wilson had made. Dicke made the short trip from Princeton to Holmdel to examine the experimental apparatus, and he realized immediately, and no doubt with some chagrin, that the BTL researchers had preempted his own experimental plans. On the other hand, Penzias and Wilson were almost certainly delighted to know that their measurement of background radiation, which they had regarded as a frustrating interference with their radio astronomy studies, constituted in itself a scientific discovery of fundamental importance.[5]

The conditions in the Penzias–Wilson experiment are represented graphically in Figure 2.12. There were no objects at ambient temperature in the field of view of the antenna, and at a frequency of 4.08 GHz, the contribution of galactic noise is negligible. The Hogg and Mumford curve at elevation angle of 90° is shown to indicate atmospheric noise of ~3°K: the value of atmospheric noise measured by Penzias and Wilson was 2.3°K. Thus, the cosmic microwave background radiation was the largest contributor to antenna noise temperature.

The "big bang" theory, that the universe had begun with a primordial explosion of a singularity in space-time ~13 billion years ago, was based in part on the observation of the astronomer Edwin Hubble in 1929. He observed that the stars and galaxies

---

[5] A. Lightman, "Radio Waves from the Big Bang," chapt. 19 in *The Discoveries* (Pantheon Books, New York, 2005) pp. 409–434.

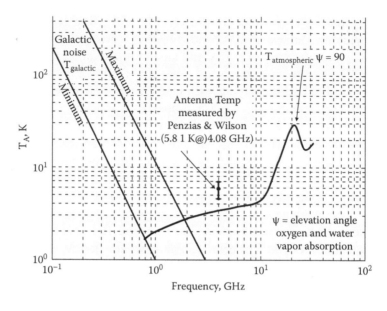

FIGURE 2.12 Experimental conditions in the Penzias–Wilson experiment.

are flying away from earth at velocities linearly proportional to their distance from the earth. The possibility of a second experimental confirmation of the big bang theory was postulated in 1946 by the physicist George Gamow, who predicted that if there had been a big bang, one residue of that primordial explosion would now take the form of a low-level background microwave radiation permeating the universe. Following soon after Gamow's prediction, Ralph Alpher and Robert Herman calculated that the temperature of this background radiation was one billion degrees above absolute zero when the universe was 100 seconds old and should have cooled over ~13 billion years to about 5°K.

Moreover, the background radiation arising from the big bang was theoretically predicted to be blackbody radiation. That prediction was confirmed in accurate measurements in the 1990s carried out using the Far-Infrared Absolute Spectrometer (FIRAS) instrument on NASA's Cosmic Background Explorer (COBE) satellite. The cosmic background radiation spectrum measured by FIRAS is shown in Figure 2.13; it corresponds closely to a 2.725°K blackbody spectrum that peaks at a frequency of 160.4 GHz, corresponding to a free space wavelength of 1.869 mm. The very close agreement between the measured data and the theoretical blackbody spectrum provides a powerful argument for the big bang theory of the origin of the universe.

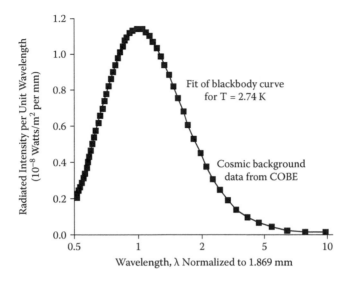

FIGURE 2.13 Spectrum of the microwave cosmic background radiation as measured by NASA.

The measurement of the microwave background radiation in the course of the painstaking experiment in 1964 carried out by Penzias and Wilson led to an ongoing series of measurements that are revealing the early history of the cosmos. Arno Penzias and Robert Wilson published the results of their excess antenna temperature measurement in the *Astrophysical Journal* in 1965, and for a likely explanation referred the reader to the paper in the same issue by R.H. Dicke, P.J.E. Peebles, P.G. Roll, and D.T. Wilkinson entitled "Cosmic Blackbody Radiation." Penzias and Wilson were awarded the Nobel Prize in Physics in 1978. Their discovery of microwave background radiation has been called the "most important scientific find of the twentieth century."[6] In 2006, the Nobel Prize in Physics was awarded to John Mather of NASA's Goddard Space Flight Center and George Smoot of the Lawrence Berkeley National Laboratory for their measurements of the spectrum of the microwave background radiation from the COBE satellite, which furnished further confirmation of the big bang theory of the origin of the universe.

---

[6] Ralph Schoenstein, "The Big Bang's Echo," May 17 2005, http://www.npr.org/templates/story.php?storyId=4655517

2.2.5 Noise Attenuation

The astute reader will have deduced from the discussions in Section 2.2.3 that when a noise from an external source with noise temperature, T_{ext}, passes through a lossy medium in which its power density is attenuated by L dB, the effective noise temperature becomes

$$T_{eff} = T_{ext} \times 10^{-L/10} + T_{med}(1 - 10^{-L/10}) \qquad (2.34)$$

where T_{med} is the noise temperature the medium would contribute if it was completely attenuating (i.e., if it was optically thick).

A relationship of the same form as Equation (2.34) also applies to the case of a lossy feeder cable between the antenna and the receiver input. If the feeder cable introduces a loss in dB of $L(C)$ and its temperature is T_c, then the equivalent noise power at the receiver input should be calculated using the equation below in place of Equation (2.16):

$$P_N = k_B \, \Delta f \, \{[(1 - \eta) \, T_o + \eta \, T_A] \, 10^{-L(C)/10}$$

$$+ T_c \, 10^{-L(C)/10} + (F - 1)T_o\} \qquad (2.35)$$

One uses Equation (2.35) in place of Equation (2.16) when a long feeder cable is used between the antenna and the receiver, for example, in a SATCOM system with the antenna mounted on the roof and with the receiver inside the house on the bottom floor. For the compact arrangement in cell phones with the antenna and receiver in close proximity, use of Equation (2.16) would usually be appropriate.

2.3 Noise in Specific Systems

2.3.1 Noise in Pagers

Pagers are used for sending an alphanumeric text message several lines in length. Such messages contain much less information than a cell phone conversation and correspondingly require much less bandwidth; not surprisingly, the portion of the electromagnetic spectrum used for paging often lies at a lower frequency than cell phone frequencies. For example, a common frequency band for paging is 138 to 174 MHz. In this band, the wavelength is about 2 m. With such a long wavelength signal being picked up by a compact handheld device, the dimensions of the receiving antenna are small compared with the wavelength, and antenna efficiency will be significantly smaller than 100%; also, at this relatively long wavelength, galactic noise will be strong. We can also assume that attenuation in the atmosphere is small at this low frequency.

The antenna will be designed to pick up signals in the horizontal plane and will be unaffected by the noise temperature of the sun or moon, so using Equations (2.18) and (2.15), we get

$$T_{cosmic} = T_{galactic} = T_o \lambda^{2.3} \qquad (2.36)$$

The total antenna temperature as given by Equation (2.17) will then have contributions from the galactic noise as given by Equation (2.36) and from the temperature of surrounding objects, $T_s = T_o$, while the big bang noise and the atmospheric noise at these frequencies are negligible. Thus, antenna temperature becomes

$$T_A = T_s + T_{cosmic} = T_o + T_o \lambda^{2.3} \qquad (2.37)$$

The equivalent noise input power is then obtained from Equation (2.16) as

$$P_N = k_B \, \Delta f \, \{[1 - \eta] \, T_o + \eta \, [1 + \lambda^{2.3}] \, T_o + [F - 1] \, T_o\}$$

$$= k_B \, \Delta f \, T_o \, [\eta \, \lambda^{2.3} + F] \qquad (2.38)$$

As a specific example, consider a pager operating at $f = 100$ MHz ($\lambda = 3$) with antenna efficiency of $\eta = 80\%$, receiver bandwidth of 5 kHz, and a receiver noise factor of $F = 2$. Using Equation (2.38), one calculates an equivalent noise input power of $P_N = 2.4 \times 10^{-16}$ Watts. In this example, galactic noise is a factor of five larger than receiver noise.

2.3.2 Noise in Cell Phones

In cell phones, antenna efficiency is nearly 100% so that noise contribution from the first term inside the square brackets in Equation (2.16) is negligible. Also, the antenna is not directed skyward and usually picks up thermal radiation from large objects such as buildings or trees at the ambient temperature, so to good approximation $T_s = T_o$, while the range of operating frequency is chosen so that both T_{cosmic} and $T_{atmospheric}$ are small compared to T_o. Thus, Equation (2.17) gives antenna temperature as approximately, $T_A = T_o$.

Then to reasonable approximation, Equation (2.16) gives the equivalent noise input power as

$$P_N = k_B \, \Delta f \, [T_o + (F - 1)T_o] = k_B \, \Delta f \, T_o \, F \qquad (2.39)$$

Comparing the expression for cell phone noise in Equation (2.39) with the expression for pager noise in Equation (2.38), it is apparent that pagers operating at lower frequency have additional noise input from galactic sources; however, in pagers the bandwidth would usually be smaller than in cell phones.

2.3.3 Noise in Millimeter-Wave SATCOM

SATCOM receiver systems are usually designed to operate in a worst-case scenario of very heavy rain (but not in the heaviest rainfall rate that is exceeded for 53 minutes in a year). This will lead to very large signal attenuation along with very large attenuation of cosmic noise. Also, the antenna is sharply focused skyward and no large surrounding objects are in the antenna's field of view. Then, the main contribution to antenna noise temperature as given by Equations (2.17) and (2.33) is the rain itself, and to good approximation

$$T_A = T_{rain} \qquad (2.40)$$

Antenna dimensions will be large compared with the wavelength to achieve strong focusing, so that antenna efficiency is close to 100%, and Equation (2.16) for the equivalent noise input power will reduce to

$$P_N = k_B \, \Delta f \, [T_{rain} + (F - 1)T_o] \qquad (2.41)$$

Equation (2.41) for noise in a millimeter-wave SATCOM system in heavy rain does not differ much from the expression for cell phone noise in Equation (2.39). However, there are very large differences between the two systems in values of the parameters that affect the received signal power, such as transmitter power, range, important path loss mechanisms, and antenna gain; these will be discussed in the chapters that follow.

Problems

2.1. Calculate the signal power that must be coupled from the antenna into the receiver in a cell phone to achieve a signal-to-noise ratio of 10 dB, given that the center frequency is 880 MHz, the noise bandwidth is 30 kHz, the amplifier noise factor is $F = 4$, and the ambient temperature is 290°K.

2.2. A pager operating at a center frequency of 50 MHz has a noise bandwidth of 5 kHz, an antenna efficiency of 50%, and a noise figure of 10 dB. Calculate the minimum signal power into the receiver if signal-to-noise ratio >5 dB is required.

2.3. Calculate the maximum acceptable path loss for a microwave link between the tops of two tall buildings operating at 10 GHz in clear weather with the following parameters:

> Transmitter power = 16 dBW
> Transmitter feeder loss = 8 dB
> Transmitter antenna gain = 20 dBi
> Receiver antenna elevation angle = 0°
> Receiver antenna gain = 13 dBi
> Receiver feeder loss = 2 dB
> Receiver noise bandwidth = 800 MHz
> Receiver noise figure = 3 dB
> Required signal-to-noise ratio >10 dB

2.4. A receiver is made up of three cascaded stages each with a gain of 10 dB and each with a noise figure of 3 dB.

   a. Calculate the overall noise figure.

   b. Calculate the overall noise figure if only the input stage degrades so that its noise figure becomes 6 dB.

   c. Calculate the overall noise figure if only the output stage degrades so that its noise figure becomes 6 dB. Discuss the reason that your answer is smaller than in Part (b).

2.5. A millimeter-wave (λ = 10 mm) SATCOM receiving antenna has a beamwidth of 1.5°. It is pointed directly at the sun but a moderately strong rain is falling, resulting in attenuation of $L(R)$ = 10 dB. Calculate the antenna temperature.

2.6. A receiving antenna in a satellite is attached to a matched receiver that has a center frequency of 160 GHz and a bandwidth of 1 GHz.

   a. Calculate the "big bang" noise power that will be received.

      b. Compare the noise power calculated in Part (a) with the additional noise power received if the moon is in the antenna beam with a filling factor of 0.3.

2.7.   For a blackbody at a temperature of 300°K, calculate the frequency at which the electromagnetic radiation spectrum will be peaked.

Bibliography

1. R.E. Collin, *Antennas and Radiowave Propagation* (McGraw-Hill, New York, 1985), 312–329.
2. S.R. Saunders, *Antennas and Propagation for Wireless Communication Systems* (John Wiley & Sons, Chichester, UK, 1999), 88–92.
3. A.A. Penzias and R.W. Wilson, "A Measurement of Excess Antenna Temperature at 4080 Mc/s," *Astrophysical Journal* 142 (1965): 419–422.
4. R.H. Dicke, P.J.E. Peebles, P.G. Roll, and D.T. Wilkinson, "Cosmic Blackbody Radiation," *Astrophysical Journal* 142 (1965): 414–422.
5. G. Gamow, "The Evolution of the Universe," *Nature* 162 (1948): 680.
6. G. Gamow, "The Origin of Elements and the Separation of Galaxies," *Physical Review* 74 (1948): 505.
7. R.A. Alpher and R. Herman, "On the Relative Abundance of the Elements," *Physical Review* 74 (1948): 1577.
8. IEEE History Center, Harry Nyquist biography, http://www.ieee.org/web/aboutus/history_center/biography/nyquist.html.
9. Wikipedia, "Black body," http://en.wikipedia.org/wiki/Black_body.

CHAPTER **3**

Antennas

Antennas are the structures used to interface between the transmitter and the wave propagation medium. The transmitter power excites currents on the antenna that, in turn, launch electromagnetic waves. Antenna gain, which is a measure of its effectiveness in focusing the electromagnetic wave power toward an intended receiver, is a key parameter in designing a wireless communication system. The connection between currents on the antenna structures and the pattern of the electromagnetic radiation they produce is described by Maxwell's equations. Thus, for the sake of intellectual integrity, we begin our description of antennas with a review of Maxwell's equations and a derivation of the inhomogeneous Helmholtz equation relating the electromagnetic fields to the current densities on an antenna. These topics are covered in some basic textbooks on electromagnetic waves (e.g., Cheng, *Field and Wave Electromagnetics*, 2nd ed., Addison-Wesley, Reading, Massachusetts), and the reader who is thoroughly familiar with these concepts might want to skip Section 3.1.

3.1 Brief Review of Electromagnetism

The modern theoretical framework for describing the interaction of electromagnetic energy with matter is known as quantum electrodynamics (QED). In 1965, Richard P. Feynman, Julian Schwinger, and Sin-Itiro Tomonaga received the Nobel Prize in Physics for developing QED. In particular, we note that Richard Feynman showed that Maxwell's equations could be derived from QED. When dealing with interactions on the scale of an atom involving only a few photons, the full QED formalism is needed.

63

On the other hand, when one deals with a very large number of photons as in the study of antennas, electromagnetic phenomena are adequately described by Maxwell's equations that relate the electric field $\mathbf{E}(\mathbf{r},t)$ and the magnetic field $\mathbf{H}(\mathbf{r},t)$ to each other and to their sources, namely, the current density $\mathbf{J}(\mathbf{r},t)$ and the charge density $\rho(\mathbf{r},t)$.

In the preceding paragraph and in the rest of this book, we represent vector quantities, such as $\mathbf{E}(\mathbf{r},t)$, by bold type. Generally, these quantities are functions of the three spatial coordinates, and of time t. In the sinusoidal steady state, which is appropriate for the topics in this book, we may work with phasor representations of the fields, E and H and phasor representations of the sources, J and ρ. The phasors are functions of the spatial coordinates but are not functions of time; the time-varying fields may be recovered from them as follows:

$$\mathbf{E}(t) = Re\left[E\ e^{j\omega t}\right] ; \mathbf{H}(t) = Re\left[H\ e^{j\omega t}\right]; \mathbf{J}(t) = Re\left[J\ e^{j\omega t}\right]; \rho(t) = Re\left[\rho e^{j\omega t}\right]$$

where $Re\ [\ ]$ denotes taking the real part of the complex quantity enclosed by the brackets.

Maxwell's equations in the form appropriate for phasor quantities are

$$\nabla \times E = -j\ \omega\ m\ H \qquad (3.1a)$$

$$\nabla \times H = j\ \omega\ \varepsilon\ E + J \qquad (3.1b)$$

$$\nabla \cdot (\varepsilon\ E) = \rho \qquad (3.1c)$$

$$\nabla \cdot (\mu\ H) = 0 \qquad (3.1d)$$

where the permeability m and the permittivity e are properties of the medium that hosts the electromagnetic fields. Equation (3.1a) is known as Faraday's law, and Equation (3.1b) is known as the Ampere–Maxwell law.

3.1.1 Maxwell's Equations and Boundary Conditions

Often one solves the set of Equations (3.1) in regions of uniform permittivity and permeability and then matches the solutions at the boundary between regions. The boundary conditions at the interface between two dielectric media are as follows:

$$E_{1,t} = E_{2,t} \tag{3.2a}$$

$$\varepsilon_1 E_{1,n} = \varepsilon_2 E_{2,n} \tag{3.2b}$$

$$H_{1,t} = H_{2,t} \tag{3.2c}$$

$$\mu_1 H_{1,n} = \mu_2 H_{2,n} \tag{3.2d}$$

where the subscripts 1 and 2 denote fields and parameters in medium #1 and medium #2, respectively; the subscripts t and n denote, respectively, the components of the fields that are tangential or normal to the boundary.

If medium #1 is a dielectric and medium #2 is a perfect conductor, from which all fields are excluded, the boundary conditions become

$$E_{1,t} = 0 \tag{3.3a}$$

$$\mathbf{a}_{n\,2} \cdot \mathbf{E}_1 = \rho_s / \varepsilon_1 \tag{3.3b}$$

$$\mathbf{a}_{n\,2} \times \mathbf{H}_1 = \mathbf{J}_s \tag{3.3c}$$

$$H_{1,n} = 0 \tag{3.3d}$$

where $\mathbf{a}_{n\,2}$ is a unit vector normal to the boundary and pointing from medium #2 into medium #1, ρ_s is the surface charge density on the conductor (in units of Coulombs/meter^2), and $\mathbf{J}_s$ is the surface current density flowing on the conductor (in units of Amperes/meter).

In principle, one can solve the set of Equations (3.1) for the fields E and H when the sources J and ρ are specified. Once the fields are determined, the time-averaged power density carried by the electromagnetic radiation is given at any point in space by

$$\mathbf{S} = \tfrac{1}{2} \, \text{Re} \, [\mathbf{E} \times \mathbf{H}^*] \qquad (3.4)$$

The point form of Maxwell's equations as represented by Equation (3.1) is due to Hertz and is general for all orthogonal curvilinear coordinate systems of which the most common are Cartesian coordinates (x,y,z); cylindrical coordinates (ρ,ϕ,z); and spherical coordinates (r,ϕ,θ). The specific form of the curl operator, $\nabla \times$, and the divergence operator, ∇. and other important differential operators for Cartesian and spherical coordinates may be found in Appendix B Cartesian coordinates and spherical coordinates are shown in Figure 3.1.

One rather simple but informative solution of Maxwell's equations known as plane waves is obtained in a homogeneous source-free region where $J = 0$ and $\rho = 0$. Then, taking the curl of Equation (3.1a), gives

$$\nabla \times \nabla \times \mathbf{E} = -\mathrm{j} \, \omega \, \mu \, \nabla \times \mathbf{H} \qquad (3.5)$$

Next, using Equation (3.1b) to express $\nabla \times \mathbf{H}$ in terms of $\mathbf{E}$ and using the identity $\nabla \times \nabla \times \mathbf{E} = \nabla(\nabla \cdot \mathbf{E}) - \nabla^2 \mathbf{E}$ gives the following equation involving only one dependent variable (viz., $\mathbf{E}$):

$$\nabla(\nabla \cdot \mathbf{E}) - \nabla^2 \mathbf{E} = \omega^2 \, \mu \, \varepsilon \, \mathbf{E} \qquad (3.6)$$

From Equation (3.1c), it may be seen that since $\rho = 0$ and the medium is homogeneous, the first term in Equation (3.6) will vanish, and we are left with

$$\nabla^2 \mathbf{E} + \mathrm{k}^2 \, \mathbf{E} = 0 \qquad (3.7)$$

where $k^2 = \omega^2 \, \mu \, \varepsilon$, and k is called the wavenumber. Equation (3.7) is called the homogeneous vector Helmholtz equation.

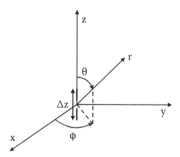

FIGURE 3.1 Cartesian and spherical coordinate systems with Hertzian dipole at the origin.

The plane wave solution of this equation is readily found in Cartesian coordinates by assuming that E is linearly polarized in the x-direction (i.e., $\boldsymbol{E} = E \, \mathbf{a}_x$) and varies spatially only in the z-direction. Then Equation (3.7) reduces to the scalar, second-order, ordinary differential equation:

$$\frac{\partial^2 E}{\partial z^2} + k^2 E = 0 \qquad (3.8)$$

that has solutions

$$E = E_1 \, e^{-jkz+j\xi} + E_2 \, e^{jkz+j\zeta} \qquad (3.9)$$

where E_1 and E_2 are amplitude constants while ξ and ζ are constants denoting phase.

The first term in Equation (3.9) represents a wave traveling in the positive z-direction while the second term represents a wave traveling in the negative z-direction. For simplicity we will set $E_2 = 0$ and only consider the wave propagating in the positive z-direction. This wave will have vector electric field

$$\boldsymbol{E} = E_1 \, e^{-jkz+j\xi} \, \mathbf{a}_x \qquad (3.10)$$

and a corresponding magnetic field, which may be found from Equations (3.1a) and (3.10) as

$$\boldsymbol{H} = (E_1/Z) \, e^{-jkz+j\xi} \, \mathbf{a}_y \qquad (3.11)$$

where $Z = \sqrt{\mu / \varepsilon}$ is the wave impedance; if the medium in which the wave propagates is free space, $Z = Z_o = \sqrt{\mu_o / \varepsilon_o} = 377$ Ohms.

Note that as the wave propagates through a distance $[z_2 - z_1]$, the phase of the electromagnetic wave is retarded by $k[z_2 - z_1]$.

The power density carried by these fields of a plane wave may be found from equations (3.4), (3.10), and (3.11) to be

$$S = \tfrac{1}{2} \, (E_1{}^2/Z) \qquad (3.12)$$

For describing antennas it is often most convenient to work with spherical coordinates with the antenna at the origin of the coordinate system and the waves propagating in the radial direction with a spherical wavefront. Nevertheless, far from the antenna, the local wavefront is almost planar, and expressions like Equation (3.12) are approximately valid.

If the medium through which the wave is propagating is slightly lossy due to a conductivity σ_c that is limited to values $\sigma_c \ll \omega\, \varepsilon$, the wave propagation may be analyzed by assuming that the medium has a complex permittivity, $\varepsilon_c = \varepsilon\,(1 - j\,\tan\delta)$ where the loss tangent, $\tan\delta = \sigma_c/(\omega\, \varepsilon)$. The power density in the electromagnetic wave will then be attenuated as it propagates by a factor $\exp(-2\, \alpha_c z)$ where the attenuation constant is $\alpha_c = \tfrac{1}{2}$ $(k\, \tan\delta)$.

3.1.2 Vector Potential and the Inhomogeneous Helmholtz Equation

When solving for the spatial pattern of the electromagnetic fields due to surface current density on a conducting structure such as an antenna, it is convenient to define the vector potential, A. In general, the divergence of the curl of any vector field is identically zero. Thus, because Equation (3.1d) specifies that the divergence of μH is zero, it is implied that μH is the curl of a vector field. We call this vector field the vector potential A; that is,

$$\mu\, H = \nabla \times A \qquad (3.13)$$

Then combining Equation (3.13) with Faraday's law, Equation (3.1a), gives

$$\nabla \times (E + j\, \omega\, A) = 0 \qquad (3.14)$$

Now, the curl of any gradient of a scalar field is identically zero. Therefore, Equation (3.14) implies that $(E + j\, \omega\, A)$ is the gradient of a scalar field; we call this scalar field $-\Phi$; that is,

$$E + j\, \omega\, A = -\nabla\, \Phi \qquad (3.15)$$

We next substitute from Equations (3.13) and (3.15) into the Ampere-Maxwell law given by Equation (3.1b) to get

$$\nabla \times \nabla \times A = \omega^2\, \mu\, \varepsilon\, A - j\, \omega\, \mu\, \varepsilon\, \nabla\, \Phi + \mu\, J \qquad (3.16)$$

We then use the vector identity $\nabla \times \nabla \times A = \nabla\nabla.\, A - \nabla^2\, A$ and note that the wavenumber $k = \omega\, (\mu\, \varepsilon)^{1/2}$ to put Equation (3.16) in the form

$$\nabla^2 A + k^2 A = -\mu J + \nabla [\nabla . A + j \omega \mu \varepsilon \Phi] \qquad (3.17)$$

When we introduced the vector potential, *A*, we specified the value of its curl in Equation (3.13); however, the value of the divergence of a vector field may be specified independently of the value of its curl. In fact, if we specify $\nabla . A$ by choosing the Lorentz Gauge, namely,

$$\nabla . A = -j \omega \mu \varepsilon \Phi \qquad (3.18),$$

the quantity in square brackets in Equation (3.17) will vanish yielding the inhomogeneous Helmholtz equation:

$$\nabla^2 A + k^2 A = -\mu J \qquad (3.19)$$

Then, by specifying the current density on an antenna structure, Equation (3.19) may be solved to find the vector potential *A*.

The magnetic field may then be calculated using Equation (3.13). The electric field may also be calculated from the vector potential; if we use Equations (3.15) and (3.18) we get

$$E = (k^2 A + \nabla\nabla . A)/(j \omega \mu \varepsilon) \qquad (3.20)$$

3.2 Radiation from a Hertzian Dipole

3.2.1 Solution of the Inhomogeneous Helmholtz Equation in the Vector Potential A

We will now solve the inhomogeneous Helmholz equation for a thin wire aligned along the *z*-axis of incremental length, Δ*z*, carrying a current, *I*. Such an incremental antenna is called a Hertzian dipole in acknowledgment of its similarity to the antenna used in the first experimental demonstration of radio waves by Heinrich Hertz. It is anticipated that the fields produced by longer wire antennas could then be found by summing the fields produced by all the increments along the antenna length.

We begin by assuming that the length of the Hertzian dipole is much smaller than the wavelength, while the radius of the Hertzian dipole is much smaller than its length. For convenience we locate the dipole at the origin of the coordinate system as shown in Figure 3.1. The current density corresponding to the current *I* on the dipole wire can then be represented by

$$J = I \, \Delta z \, a_z \, \delta \, (r) \tag{3.21}$$

where $\delta(\mathbf{r})$ is the Dirac delta function located at the origin.

The Dirac delta function has the property that

$$\int_V \delta \, (\mathbf{r}) \, dV = 1$$

if the origin ($\mathbf{r} = 0$) is in the volume of integration, and

$$\int_V \delta \, (\mathbf{r}) \, dV = 0$$

if the origin ($\mathbf{r} = 0$) is not in the volume of integration.

It should be clear from the integral expressions above that the Dirac delta function has dimensions of (volume)$^{-1}$ and therefore the expression for J in Equation (3.21) has dimensions of Amps/meter^2 as required.

Substituting Equation (3.21) into the inhomogeneous Helmholtz equation, Equation (3.19), and noting that in free space $\mu = \mu_o = 4\pi \times 10^{-7}$ Henries/meter, we obtain

$$\nabla^2 \mathbf{A} + k^2 \mathbf{A} = -\mu_o \, I \, \Delta z \, \delta(\mathbf{r}) \, \mathbf{a}_z \tag{3.22}$$

Next, because the direction of the vector on the right-hand side is $\mathbf{a}_z$, we will assume that on the left-hand side,

$$\mathbf{A} = A \, \mathbf{a}_z \tag{3.23}$$

which allows one to convert Equation (3.22) into a scalar equation:

$$\nabla^2 A + k^2 \, A = -\mu_o \, I \, \Delta z \, \delta(\mathbf{r}) \tag{3.24}$$

As an additional simplification, we note that the source term on the right-hand side of Equation (3.23) does not depend on f or q, and accordingly, we will assume that A is only a function of the r coordinate. This assumption allows one to write Equation (3.24) as an ordinary differential equation:

$$\frac{1}{r^2} \frac{d}{dr} (r^2 \frac{dA}{dr}) + k^2 A = -\mu_o \, I \, \Delta z \, \delta(\mathbf{r}) \tag{3.25}$$

At all points in space, except for **r** = 0 where the Hertzian dipole antenna is located, Equation (3.25) becomes simply

$$\frac{1}{r^2}\frac{d}{dr}(r^2\frac{dA}{dr}) + k^2 A = 0, (r \neq 0) \tag{3.26}$$

The solutions to this second-order, ordinary, differential equation are

$$A = C_1\frac{e^{-jkr}}{r} + C_2\frac{e^{jkr}}{r}, (r \neq 0) \tag{3.27}$$

which may be verified by substitution; C_1 and C_2 are constants of integration.

In Equation (3.27), the first term on the right-hand side represents a wave propagating radially outward from the origin. The second term on the right-hand side represents a wave propagating radially inward. Because we are interested only in outward propagating waves produced by the current on the Hertzian dipole, we set $C_2 = 0$, so that Equation (3. 27) becomes

$$A = C_1\frac{e^{-jkr}}{r}, (r \neq 0) \tag{3.28}$$

To evaluate C_1, integrate Equation (3.24) over the volume of a small sphere centered at the origin and having radius r_o to get

$$\int_V (\nabla^2 A + k^2 A)\, dV = -\mu_o I\, dz \tag{3.29}$$

Now

$$\int_V k^2 A\, dV = 4\pi \int_0^{r_o} A\, k^2\, r^2\, dr \rightarrow 0 \text{ as } r_o \rightarrow 0$$

Therefore, Equation (3.29) becomes

$$\int_V \nabla^2 A\, dV = \int_V \nabla.\nabla A\, dV = -\mu_o I\, dz \tag{3.30}$$

Then, using the divergence theorem, the left-hand side of Equation (3.30) can be converted from a volume integral into a surface integral giving

$$\int_S \nabla A . \, \mathbf{a}_r \, ds = 4\pi \, r_o^2 \frac{dA}{dr} = -\mu_o I \, \Delta z \tag{3.31}$$

with dA/dr being evaluated at $r = r_o$.

Then, substituting for A from Equation (3.28) and performing the differentiation indicated in Equation (3.31),

$$4\pi \, r_o^2 \, C_1 \left(\frac{-jke^{-jkr}}{r} - \frac{e^{-jkr}}{r^2} \right) = -\mu_o \, I \Delta z \tag{3.32}$$

Next, one makes the substitution of r_o for r indicated in Equation (3.31), and takes the limit as $r_o \to 0$ to obtain

$$4\pi \, C_1 = \mu_o I \, \Delta z \tag{3.33}$$

Substituting the value of C_1 as given by Equation (3.33) into Equations (3.23) and (3.28) gives the sought for expression for the vector potential:

$$A = \frac{1}{4\pi} \mu_o \, I \, \Delta z \, \frac{e^{-jkr}}{r} \, \mathbf{a}_z$$

$$= \frac{1}{4\pi} \mu_o \, I \Delta z \, \frac{e^{-jkr}}{r} \, [\cos\theta \, \mathbf{a}_r - \sin\theta \, \mathbf{a}_\theta] \tag{3.34}$$

The magnetic and electric fields in the radially outward propagating wave that are produced by the current on the Hertzian dipole may now be calculated by substituting the expression for the vector potential given by Equation (3.34) into Equations (3.13) and (3.20).

3.2.2 Near Fields and Far Fields of a Hertzian Dipole

By the methods described above we obtain the expressions for the vector phasors representing the magnetic and electric fields of the waves produced by an alternating current represented by the phasor I on a thin wire of incremental length Δz aligned with the z axis; such a small wire antenna is called a Hertzian dipole. These expressions for the fields in spherical coordinates are as follows:

$$\mathbf{H} = \mathbf{a}_\phi \, j \, (4\pi)^{-1} \, I \, \Delta z \, e^{-jkr} \, k^2 \left[\frac{1}{kr} - \frac{1}{(kr)^2} \right] \sin\theta \qquad (3.35)$$

$$\mathbf{E} = -\mathbf{a}_r \, j \, (2\pi)^{-1} \, I \, \Delta z \, e^{-jkr} \, Z_o \, k^2 \left[\frac{j}{(kr)^2} + \frac{1}{(kr)^3} \right] \cos\theta$$

$$+ \, \mathbf{a}_\theta \, j \, (4\pi)^{-1} \, I \, \Delta z \, e^{-jkr} \, Z_o \, k^2 \left[-\frac{1}{kr} + \frac{j}{(kr)^2} + \frac{1}{(kr)^3} \right] \sin\theta$$

$$(3.36)$$

where $Z_o = (\mu_o/\varepsilon_o)^{1/2} = 377$ Ohms is the wave impedance of free space.

Of special interest in wireless communications are the radiation fields far from the antenna (i.e., the "far fields"). For an antenna with dimensions much smaller than a wavelength, the far fields may be found by taking the limit as

$$kr \gg 1 \qquad (3.37)$$

Applying this limit to Equations (3.35) and (3.36), we keep only those terms that fall off with radial distance as $(kr)^{-1}$, and obtain the following expressions for the far fields:

$$\boldsymbol{H} = \mathbf{a}_\phi \, j \, (4\pi)^{-1} \, I \, \Delta z \, e^{-jkr} \, k \sin\theta/r \qquad (3.38)$$

$$\boldsymbol{E} = \mathbf{a}_\theta \, j \, (4\pi)^{-1} \, I \, \Delta z \, e^{-jkr} \, Z_o \, k \sin\theta/r \qquad (3.39)$$

Note that in the far field, $\boldsymbol{H}$ and $\boldsymbol{E}$ are in phase, the magnitude of $\boldsymbol{E}$ is Z_o times the magnitude of $\boldsymbol{H}$, and $\boldsymbol{E} \times \boldsymbol{H}$ is in the $\mathbf{a}_r$ direction (i.e., the direction of wave propagation is radially outward in spherical coordinates).

These properties are common to the far fields of all antennas, but for antennas with dimensions that are not small compared with the wavelength, more stringent conditions are obtained for the far field than the one given by Equation (3.37). Three far field conditions, all of which must be satisfied for the fields of the radiating wave to be dominant over the reactive fields, are

$$r > 2 \, \Lambda^2 / \lambda \qquad (3.40a)$$

and

$$r \gg \Lambda \qquad (3.40\text{b})$$

and

$$r \gg \lambda \qquad (3.40\text{c})$$

where L is the largest dimension of the antenna.

The terms in Equations (3.35) and (3.36), which decrease with distance from the antenna more rapidly than $(kr)^{-1}$, are part of the "near field" and represent reactive stored energy in the region near the antenna. The near field region is also called the Fresnel zone, while the far field region is called the Fraunhofer zone. In the far field region of a Hertzian dipole, only the H_ϕ and E_θ remain as given by Equations (3.38) and (3.39); these represent radiated power that is propagating away from the antenna in the radial direction.

3.2.3 Basic Antenna Parameters

A radiation pattern of an antenna is a graph displaying the relative, Fraunhofer zone, field strength as a function of direction at a fixed distance from the antenna. For the Hertzian dipole we have

$$|E_\theta|/|E_\theta|_{\max} = \sin\theta \qquad (3.41)$$

This pattern is plotted in Figure 3.2 in both the *E*-plane and the *H*-plane. The *E*-plane is a plane in which the Fraunhofer zone *E* field lies (e.g., the x,z plane or the y,z plane). The *H*-plane is a plane in which the Fraunhofer zone *H* field lies (e.g., the x,y plane). Note that the radiation pattern of a Hertzian dipole is omnidirectional (i.e., it does not vary with direction in the horizontal x,y plane). A three-dimensional plot of the radiation field of a Hertzian dipole is presented in Figure 3.3.

The radiation pattern of a Hertzian dipole is especially simple in its form. In general an antenna radiation pattern will be more complicated. A generic example of a more complicated *H*-plane pattern is shown in Figure 3.4. It has a number of salient features that are worthwhile defining. The lobe in the pattern that contains the maximum value of normalized electric field is called the *main lobe*. In this lobe, the angular distance between the points where the field strength falls to its maximum value minus 3 dB is called

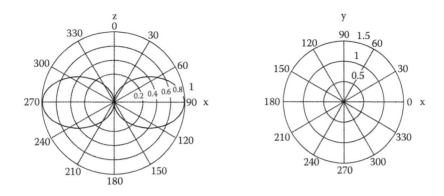

FIGURE 3.2 Radiation pattern of a Hertzian dipole. E-plane plot shown on the left. H-plane plot shown on the right.

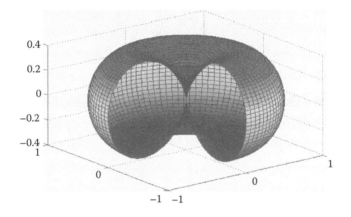

FIGURE 3.3 Three-dimensional presentation of radiation pattern of a Hertzian dipole.

the *beamwidth*. [The 3 dB beamwidth of the Hertzian dipole radiation pattern is $BW_\theta = 90°$, which may be readily checked by considering that $(\sin 45°)^2 = (\sin 135°)^2 = 0.5$.] The lobe displaced from the main lobe by 180° is called the *back lobe*. Other lobes in the pattern are known as *side lobes*.

From the beamwidth of the omnidirectional radiation pattern of the Hertzian dipole, we can calculate antenna directivity using Equation (1.13):

$$D = 102°/[\ BW_\theta(deg.) - 0.0027\ \{BW_\theta(deg.)\}^2]$$

$$= 102/[90 - 0.0027\ (90)^2] = 1.5$$

Directivity will be discussed further in the following section.

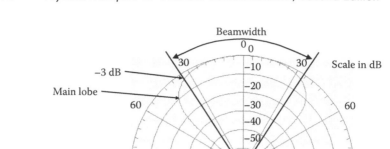

FIGURE 3.4 General form of an antenna radiation pattern.

3.2.4 Directive Gain, D(φ,θ); Directivity, D; and Gain, G

The pattern of radiated power is described by a parameter known as the directive gain defined by

$$D(\phi,\theta) = S(\phi,\theta)/S_I \tag{3.42}$$

where $S(\phi,\theta)$ is the magnitude of the vector representing the time average radiated power density, S, which is given by Equation (3.4).

S_I is the magnitude of the power density that would be obtained from an isotropic radiator:

$$S_I = P_r/(4\pi r^2) \tag{3.43}$$

Combining Equations (3.42) and (3.43) gives

$$D(\phi,\theta) = 4\pi r^2 \, S(\phi,\theta)/P_r \tag{3.44}$$

For a Hertzian dipole with radiation fields given by Equations (3.38) and (3.39), we calculate

$$S = a_r(I \, \Delta z)^2 \, Z_o \, k^2 \sin^2\theta/(32 \, \pi r^2) = a_r \, S(\phi,\theta) \tag{3.45}$$

while the total average radiated power is given by

$$P_r = \oiint S \cdot ds \qquad (3.46)$$

where $ds = a_r \, r^2 \sin \theta \, d\phi \, d\theta$, and the indicated integration is over the closed surface of a sphere centered at the origin at any fixed radial distance in the Fraunhofer zone.

Then, substituting from Equation (3.45) into Equation (3.46) and performing the indicated integrations gives

$$P_r = (I \, \Delta z)^2 \, Z_o \, k^2/(12\pi) \qquad (3.47)$$

Then substituting from Equations (3.45) and (3.47) into Equation (3.44) gives the following expression for the directive gain of a Hertzian dipole

$$D(\phi,\theta) = 1.5 \sin^2\theta \qquad (3.48)$$

This expression gives the radiated power density of a Hertzian dipole in any direction normalized to the power density that would result from an isotropic radiator.

Antennas are often characterized by their maximum value of directive gain that is called the directivity and is denoted by the symbol D. From Equation (3.44) it may be seen that

$$D = 4\pi r^2 \, S_{max}/P_r \qquad (3.49)$$

where S_{max} is the maximum value of the radiation power density $S(\phi,\theta)$.

For a Hertzian dipole, the maximum value of directive gain occurs in the horizontal plane ($\theta = \pi/2$) where one can deduce from Equation (3.38) that $D = 1.5$ or equivalently $D = 1.76$ dBi. This value is in agreement with the value calculated using Equation (1.13) as discussed in Section 3.2.3.

An even more ubiquitous parameter used to characterize antenna performance is gain, G, which is simply the product of Ohmic efficiency and directivity. In general,

$$G = \eta \, D \qquad (3.50)$$

Gain of an antenna is a measure of the maximum radiated power density produced with the aid of that antenna in a preferred direction while accounting for antenna losses normalized to the power density that would have been achieved with a lossless isotropic radiator.

3.2.5 Radiation Resistance of a Hertzian Dipole Antenna

The analytical results that have been obtained above for a Hertzian dipole were obtained assuming antenna length «l and with uniform current magnitude along the antenna length. We are now able to calculate the radiation resistance of an antenna that satisfies the Hertzian dipole conditions. The radiated power for a Hertzian dipole as given by Equation (3.47) may be set equal to the general expression for radiated power in terms of the radiation resistance:

$$P_r = \tfrac{1}{2}\,1I_i|^2\,R_r \tag{3.51}$$

[Note that Equation (3.51) is similar to Equation (2.3), but we are now using phasor notation for antenna input current: $1I_i1 = I_o$.]

Setting Equation (3.51) equal to Equation (3.47) gives

$$R_r = 2\,(\Delta z)^2\,Z_o\,k^2/12\,\pi$$

or

$$R_r = 80\,\pi^2\,(\Delta z/\lambda)^2\,\text{Ohms} \tag{3.52}$$

where we have used $Z_o = 120\,\pi$ Ohms and $k = 2\pi/\lambda$.

3.2.6 Electrically Short Dipole Antenna (Length « λ)

A more realistic model of a wire dipole antenna is one in which the current along the wire goes to zero at the ends. For a dipole of half-length h, it is appropriate to model the current as

$$I(z) = I_i\,\frac{\sin\,[k(h - |z|)]}{\sin\,(k\,h)},\ -h < z < h, \tag{3.53}$$

where I_i is the current at the central feedpoint of the antenna (i.e., at z = 0).

For an electrically short antenna with h « λ or kh « 1, Equation (3.53) becomes

$$I\,(z) = I_i\,[1 - (|z|\,/h)],\ -h < z < h \tag{3.54}$$

Thus, the current distribution along the dipole is triangular as shown in Figure 3.5.

If we also assume that the antenna is sufficiently short that in the far field it effectively appears to be a point so the factor [$\sin\theta$ e^{-jkr}/r] remains constant for any position along the antenna, we can see from our analysis of the Hertzian dipole and specifically from Equation (3.39) that the contribution to the electric field from an incremental length along the antenna, dz, at location z is

$$dE_\theta(z) = j(4\pi)^{-1} I(z) \, dz \, Z_0 \, k \sin\theta \, e^{-jkr}/r \tag{3.55}$$

Then, summing up the contributions to the field from all the incremental lengths along the antenna, we get

$$E_\theta = [j \, I_i (4\pi)^{-1} \, Z_0 k \sin\theta \, e^{-jkr}/r] \times \left[2 \int_0^h [1 - (z/h)]dz \right]$$

or

$$E_\theta = j \, I_i h \, (4\pi)^{-1} \, Z_0 \, k \sin\theta \, e^{-jkr}/r \tag{3.56}$$

which is of the same form as the field produced by a Hertzian dipole, except that the current at the feedpoint I_i replaces the uniform current I, and the half length of the short dipole h replaces the full length of the Hertzian dipole Δz.

Thus, if we now define the radiation resistance in terms of the current at the feedpoint, so that the radiated power from the short dipole is

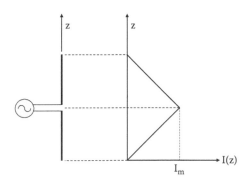

FIGURE 3.5 Triangular current distribution on an electrically short dipole antenna.

$$P_r = \tfrac{1}{2} |I_i|^2 R_r \qquad (3.57)$$

we can obtain an expression for the radiation resistance similar in form to Equation (3.52):

$$R_r = 80 \, \pi^2 (h/\lambda)^2 = 20 \, \pi^2 (\ell/\lambda)^2 \qquad (3.58)$$

where $\ell = 2\,h$ is the full length of the dipole antenna.

To get the Ohmic power loss in the short dipole, we note that

$$P_o = \tfrac{1}{2} \int_{-h}^{h} [|I(z)|^2 \, R_s /(2\pi \, r_w)] \, dz \qquad (3.59)$$

where R_s is the surface resistance of the metallic antenna wire, and r_w is the wire radius.

Then, substituting the expression for $I(z)$ in Equation (3.54) into Equation (3.59) gives

$$P_o = |I_i|^2 \, R_s \, (2\pi \, r_w)^{-1} \int_0^h [1 - (z/h)]^2 \, dz \qquad (3.60)$$

To evaluate the integral in Equation (3.60), we make the change of variable $\xi = [1 - (z/h)]$. Then, Equation (3.60) becomes

$$P_o = |I_i|^2 \, R_s h \, (2\pi \, r_w)^{-1} \int_0^1 \xi^2 d\xi$$

$$= |I_i|^2 \, R_s h \, (2\pi \, r_w)^{-1} \, (\tfrac{1}{3}) \qquad (3.61)$$

The Ohmic resistance may be defined in terms of the Ohmic power loss and the feedpoint current as

$$P_o = \tfrac{1}{2} |I_i|^2 R_o \qquad (3.62)$$

Then, comparing Equations (3.61) and (3.62) we see that the Ohmic resistance may be expressed as

$$R_o = (1/3) \, R_s \, \ell \, (2\pi \, r_w)^{-1} \qquad (3.63)$$

We can now use the expression for radiation resistance in Equation (3.58) and the expression for Ohmic resistance in Equation (3.63) to calculate the efficiency of a short dipole antenna.

As an example, consider the case of a 2 cm long copper dipole that has a wire diameter of 1 mm being employed to receive a narrowband pager signal centered at a frequency of 150 MHz (λ = 2 m). The radiation resistance is

$$R_r = 20 \pi^2 (\ell/\lambda)^2$$

$$= 20 \pi^2 (0.02/2)^2$$

$$= 0.0197 \text{ Ohms}$$

To calculate the Ohmic resistance, we begin by evaluating the surface resistance:

$$R_s = (\pi f \mu_o/\sigma_{cu})^{1/2}$$

$$= [\pi \times 1.5 \times 10^8 \times 4\pi \times 10^{-7}/(5.8 \times 10^7)]^{1/2}$$

$$= 0.00319 \text{ Ohms}$$

Then,

$$R_o = (1/3) R_s \ell (2\pi r_w)^{-1}$$

$$= (1/3) \times 0.00319 \times 0.02 \times /(2\pi \times 0.0005)$$

$$= 0.00677 \text{ Ohms}$$

The antenna Ohmic efficiency then is

$$\eta = R_r/(R_r + R_o)$$

$$= 74\%$$

Note that radiation resistance varies as $(\ell/\lambda)^2$ while Ohmic resistance is linearly proportional to ℓ, so that as antenna length becomes a larger fraction of the wavelength, efficiency will approach 100%.

Because the antenna is electrically short, implying that the factor [$\sin\theta \ e^{-jkr}/r$] is constant for each incremental current position along the antenna, the radiation pattern of the short dipole will be the same as for the Hertzian dipole and its directivity $D = 1.5$.

Gain is the product of directivity and efficiency, and for the example above,

$$G = 1.5 \times 0.774 = 1.16$$

or

$$G(dBi) = 0.65 \text{ dBi}$$

3.2.7 Small Loop Antennas

Small loop antennas have a potential advantage compared with the small electric dipole antenna described above in that their radiation resistance can be increased by using multiple turns or by using a ferrite core. The small loop antenna may be thought of as a "magnetic dipole" having a magnetic field pattern that is similar to the electric field pattern produced by an electric dipole. The fields are sketched in Figure 3.6.

To analyze the small loop antenna, we will model it as a square loop in the x-y plane with the length of each side $= \ell$ and with area $\Sigma = \ell^2$. The circulating current is I as shown in Figure 3.7. (Our analysis is similar to that in Sturtzman and Thiele, *Antenna Theory and Design*, John Wiley & Sons, New York, 1981.)

Referring to the form of the vector potential displayed in Equation (3.34), we can state that the two sides of the loop that are parallel to the x-axis have a total vector potential that is x-directed with magnitude

$$A_x = \mu_o \, I \, \ell \, \{[\exp(-jkR_1)/(4\pi R_1)] - [\exp(-jkR_3)/(4\pi R_3)]\}$$

$$(3.64a)$$

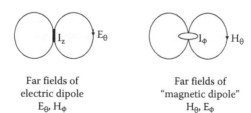

Far fields of
electric dipole
E_θ, H_ϕ

Far fields of
"magnetic dipole"
H_θ, E_ϕ

FIGURE 3.6 Sketches to compare fields of an electric dipole and a "magnetic dipole."

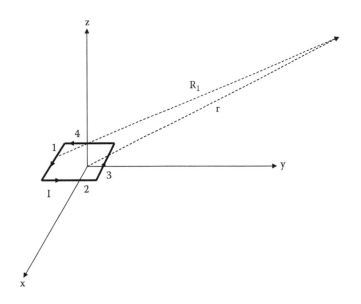

FIGURE 3.7 Small square loop antenna.

Similarly, the two sides parallel to the y-axis have a total vector potential that is y directed with magnitude

$$A_y = \mu_o \, I \, \ell \, \{[\exp(-jkR_2)/(4\pi \, R_2)] - [\exp(-jkR_4)/(4\pi \, R_4)]\}$$

$$(3.64b)$$

Next, we make the far field approximation that

$$R_1 \approx R_2 \approx R_3 \approx R_4 \approx r \qquad (3.65)$$

However, small differences between these distances must be retained when calculating phase. For phase calculations, we take

$$R_1 = r + \frac{1}{2}\ell \sin\theta \sin\phi \qquad (3.66a)$$

$$R_2 = r - \frac{1}{2}\ell \sin\theta \sin\phi \qquad (3.66b)$$

$$R_3 = r - \frac{1}{2}\ell \sin\theta \sin\phi \qquad (3.66c)$$

$$R_4 = r + \frac{1}{2}\ell \sin\theta \sin\phi \qquad (3.66d)$$

Substituting the approximations in Equations (3.65) and (3.66) into the expressions for the vector potential magnitudes in Equation (3.64), we get

$$A_x = \mu_o \, I\ell \left[e^{-jkr}/(4\pi r) \right]$$

$$\left[\exp\left(-j\tfrac{1}{2}k\ell \sin\theta \, \sin\phi\right) - \exp\left(j\tfrac{1}{2}k\ell \sin\theta \, \sin\phi\right) \right]$$

$$= -2j\mu_o \, I\ell \, [e^{-jkr}/(4\pi r)] \sin(\tfrac{1}{2}k\ell \sin\theta \, \sin\phi) \qquad (3.67a)$$

$$A_y = \mu_o \, I\ell \left[e^{-jkr}/(4\pi r) \right]$$

$$\left[\exp\left(j\tfrac{1}{2}k\ell \sin\theta \cos\phi\right) - \exp\left(-j\tfrac{1}{2}k\ell \sin\theta \cos\phi\right) \right]$$

$$= 2j \, \mu_o \, I\ell \, [e^{-jkr}/(4\pi r)] \sin(\tfrac{1}{2}k\ell \sin\theta \cos\phi) \qquad (3.67b)$$

Now, for a small loop, $k\ell \ll 1$, so that the sine functions in Equation (3.67) may be approximated by their arguments. This yields the following expression for the vector potential:

$$\mathbf{A} = A_x \, \mathbf{a}_x + A_y \, \mathbf{a}_y$$

$$= j\,k\,\ell^2\,\mu_o\,I\,[e^{-jkr}/(4\pi r)]\sin\theta\,[-\sin\phi\,\mathbf{a}_x + \cos\phi\,\mathbf{a}_y]$$

$$= [j\,k\,\Sigma\,\mu_o\,I\sin\theta\,[e^{-jkr}/(4\pi r)]\,\mathbf{a}_\phi \qquad (3.68)$$

Note that for the form of the vector potential $\mathbf{A}$ given by Equation (3.68), $\nabla \bullet \mathbf{A} = 0$.

Then, Equation (3.20) that allows one to calculate the electric field reduces to

$$\mathbf{E} = k^2\,\mathbf{A}/(j\,\omega\,\mu_o\,\varepsilon_o) = Z_o\,k^2\,\Sigma\,I\sin\theta\,[e^{-jkr}/(4\pi r)]\,\mathbf{a}_\phi \quad (3.69)$$

The magnetic field produced by the current in the small loop is then

$$\mathbf{H} = (1/Z_o)\,\mathbf{a}_r \times \mathbf{E} = -k^2\,\Sigma\,I\sin\theta\,[e^{-jkr}/(4\pi r)]\,\mathbf{a}_\theta \quad (3.70)$$

Equations (3.69) and (3.70) that give the fields for the small loop antenna can then be used to calculate the average radiated power density as

$$S = \tfrac{1}{2} E \times H^* = (|I| \, k^2 \Sigma \, \sin\theta)^2 \, Z_o/(32\pi^2 r^2) a_r \qquad (3.71)$$

The total radiated power is found by integrating **S** over the surface of a sphere surrounding the antenna:

$$P_r = \iint S \bullet a_r \, r^2 \sin\theta \, d\theta \, d\phi = (|I| \, k^2 \, \Sigma \,)^2 \, Z_o/12\pi \qquad (3.72)$$

Because the total radiated power is also equal to $\tfrac{1}{2}|I|^2 R_r$, we can deduce from Equation (3.72) that the radiation resistance for the small loop antenna is

$$R_r = 20 \, (k^2 \, \Sigma)^2 = 31{,}200 \, (\Sigma/\lambda^2)^2 \, \text{Ohms} \qquad (3.73)$$

For a small circular loop of radius r_1 and wire radius r_w, the Ohmic resistance is

$$R_o = R_s(r_1/r_w) \, \text{Ohms} \qquad (3.74)$$

and the inductance is

$$L = r_1 \mu_o \, [\ln(8 \, r_1/r_w) - 1.75)] \, \text{Henries} \qquad (3.75)$$

Design Calculation for a Small Loop Antenna

Given: Loop circumference $2\pi \, r_1 = 0.2 \, \lambda$, wire is copper of radius $r_w = 0.001 \, \lambda$, and $f = 3$ MHz
Calculate the following parameters:

Radiation resistance, R_r
Ohmmic resistance, R_o
Efficiency, η
Reactive impedance, $X = \omega L$

Solution:

From Equation (3.73), radiation resistance

$$R_r = 31{,}200 \, (\pi \, r_1^2/\lambda^2)^2 = 31{,}200 \, (0.01/\pi)^2 = 0.316 \, \text{Ohms}$$

From Equation (3.74), Ohmic resistance

$$R_o = R_s \, (r_l/r_w) = [\pi f \mu_o/(2 \, \sigma_{cu})] \, [100/\pi] = [4.52 \times 10^{-4}][31.8]$$
$$= 1.44 \times 10^{-2} \text{ Ohms}$$

Then the antenna efficiency is

$$\eta = R_r/(R_r + R_o) = 0.316/0.330 = 95.8\%$$

Finally using Equation (3.75) we calculate the reactive impedance as

$$X = 2\pi f L = 2 \, \pi \times 3 \times 10^6 \times r_l \, \mu_o \, [\ln(8 \, r_l/r_w) - 1.75)]$$
$$= 858 \text{ Ohms}$$

Thus, we see that the efficiency of the small loop antenna is acceptably large. However, the inductive reactance is much larger than the radiation resistance that will make it difficult to compensate for this reactance especially over any appreciable frequency band.

One solution to this difficulty would be to raise the radiation resistance by using multiple turns or a ferrite core. For a loop with n turns and a ferrite core with relative permeability μ_r, the expression for radiation resistance becomes

$$R_r = 31,200 \, (n \, \mu_r \, \Sigma/ \, \lambda^2)^2 \text{ Ohms} \qquad (3.76)$$

Using Equation (3.76), for the small loop dimensions in the example above but now specifying five turns and a ferrite core with $\mu_r = 10$, we find $R_r = 789$ Ohms.

3.3 Receiving Antennas, Polarization, and Aperture Antennas

3.3.1 Universal Relationship between Gain and Effective Area

As we have seen, a transmitting antenna is characterized by its gain, G. A receiving antenna will be characterized by its effective area, A_e. Effective area may be defined by considering an incoming signal with incident power density, S_i. The maximum power delivered to a matched load connected to the antenna is

$$P_i = A_e \, S_i \qquad (3.77)$$

where we have assumed that the directions of the incoming wave-vector and the incoming wave electric field are optimum for maximum received power.

Often the same antenna is connected to both a transmitter and a receiver (the transmitter–receiver combination is often referred to as a *transceiver*). One would expect the antenna gain, G, when used for transmission, to be related to the antenna's effective area, A_e, when used for receiving. Such a relationship does exist and is of the same form for all antennas.

We will find this relationship by conducting a simple gedanken experiment using the situation in Figure 3.8 where two antennas are shown separated by a distance d. Each antenna is connected to its own transceiver. First consider the case in which antenna #1 is transmitting and antenna #2 is receiving. The incoming power density at antenna #2 is

$$S_{i2} = P_{t1} \, G_1/(4 \, \pi \, d^2) \qquad (3.78)$$

where P_{t1} is the output power of transmitter #1, and G_1 is the gain of antenna #1.

The power into receiver #2 assuming matched input impedance and optimized antenna orientation is

$$P_{r2} = A_{e2} \, S_{i2}$$

$$= P_{t1} \, A_{e2} \, G_1/(4 \, \pi \, d^2) \qquad (3.79)$$

where we have used Equation (3.78).

Equation (3.78) may be rearranged to give the ratio of received power to transmitter power:

$$P_{r2}/P_{t1} = A_{e2} \, G_1/(4 \, \pi \, d^2) \qquad (3.80)$$

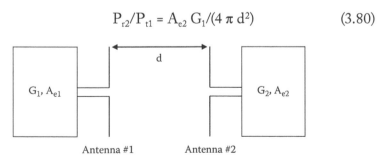

FIGURE 3.8 Geometry for gedanken experiment to find the relationship between G and A_e Each antenna is attached to its own transceiver. Distance between antennas is d, which is assumed large enough to satisfy far field conditions (not to scale).

Next, consider the case in which antenna #2 is transmitting and antenna #1 is receiving. Then by a similar reasoning process to that presented in the previous paragraph,

$$P_{r1}/P_{t2} = A_{e1}\, G_2/(4\,\pi\, d^2) \tag{3.81}$$

Because our system is linear, we may invoke reciprocity (i.e., the fraction of transmitter power received when transmitting from antenna #1 to antenna #2 will equal the fraction of transmitter power received when transmitting from antenna #2 to antenna #1). In other words,

$$P_{r2}/P_{t1} = P_{r1}/P_{t2} \tag{3.82}$$

Then substituting into Equation (3.82) from Equations (3.80) and (3.81) gives

$$A_{e1}/G_1 = A_{e2}/G_2 \tag{3.83}$$

Because nothing was specified about the nature of antenna #1 or about the nature of antenna #2, Equation (3.83) expresses the fact that A_e/G is a universal constant applicable to any antenna.

To find the value of this universal constant we need only consider one antenna. The antenna that we choose to consider is a Hertzian dipole without Ohmic losses. We determined in Section 3.2.4 that as given by Equations (3.49) and (3.50), the gain of such an antenna is

$$G = 1.5 \tag{3.84}$$

Also, the radiation resistance as given in Section 3.2.5 by Equation (3.52) is

$$R_r = 80\,\pi^2\, (\Delta z/\lambda)^2 \tag{3.85}$$

Now consider a plane wave, with electric field, E_i, incident on such an antenna with the direction of the field aligned with the antenna wires. The power density in the incoming wave is

$$S_i = \tfrac{1}{2}\, |E_i|^2/Z_o \tag{3.86}$$

and the voltage generated in the antenna is then $V_g = E_i\, \Delta z$.

The equivalent circuit of the Hertzian dipole receiving antenna is shown in Figure 3.9. The antenna impedance $Z_r = R_r + j X_r$ and the receiver input impedance is matched to the antenna impedance so $Z_i = Z_r^*$. The receiver input power is then

$$P_i = \tfrac{1}{2}\, \text{Re}\,(V_i\, I_i^*)$$

$$= \tfrac{1}{2}\, \text{Re}\,[\{V_g\, Z_r^*/(Z_r + Z_r^*)\}\,\{V_g/(Z_r + Z_r^*)\}^*]$$

$$= (|E_i|\Delta z)^2/(8\, R_r) \qquad (3.87)$$

Recall from Equation (3.77) that

$$A_e = P_i/S_i = Z_o\, \Delta z^2/4\, R_r \qquad (3.88)$$

where we have used Equations (3.86) and (3.87).

Then substituting into Equation (3.88) from Equations (3.85), and recalling that $Z_o = 120\,\pi$ Ohms gives

$$A_e = (\lambda^2/4\pi)\,(3/2) \qquad (3.89)$$

Next, using Equation (3.83), we have

$$A_e/G = (\lambda^2/4\pi) \qquad (3.90)$$

Equation (3.90) is the universal relationship between effective area and gain (or between effective area and directivity because $G = \eta D$) that holds true for any antenna.

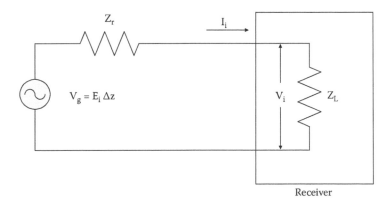

FIGURE 3.9   Equivalent circuit of Hertzian dipole receiving antenna.

3.3.2 Friis Transmission Formula

We are now in a position to evaluate the ratio of received power to transmitter power for propagation in free space over a distance, d, where the signal is simply expanding with a spherical wavefront without interference. In that case, the maximum power density radiated by a transmitter with power, P_t, cable loss, L_t, and antenna gain, G_t, is

$$S = \frac{P_t \, G_t}{L_t \, 4\pi \, d^2} \qquad (3.91)$$

The maximum signal power into the receiver, with matched input impedance, attached to a receiving antenna with effective area A_e, cable loss L_r, and gain G_r is

$$P_s = S \, A_e / L_r \;\; = \;\; \frac{P_t \, G_t}{L_t \, 4\pi \, r^2} \, \frac{(\lambda^2/4\pi) \, G_r}{L_r} \qquad (3.92)$$

where we have used Equation (3.7) and Equation (3.91). Also, we have used the symbol, r, for the range (i.e., the distance between transmitting and receiving antennas).

From Equation (3.92) it is readily seen that when cable losses are negligible, the ratio of received power to transmitter power is

$$P_s/P_t = (16 \, \pi^2)^{-1} \, (\lambda/r)^2 \, G_t \, G_r \qquad (3.93)$$

Equation (3.93) is the well-known Friis transmission formula.

3.3.3 Polarization Mismatch

The Friis transmission formula gives the maximum fraction of transmitter power that could possibly be received. Many factors will reduce this fraction such as absorption and scattering of the wave; propagation models that include these processes will be considered in Chapter 5. One important factor, which we will consider in this section, is mismatch between the polarization of the incoming wave and the polarization of the receiving antenna.

The polarization of an electromagnetic wave is the direction of its electric field. We will designate the polarization by the unit vector **h**. For example, a wave that was linearly polarized in the θ direction would have $\mathbf{h} = \mathbf{a}_\theta$, while a wave linearly polarized in the ϕ direction would have $\mathbf{h} = \mathbf{a}_\phi$.

Another polarization class of technological interest is circular polarization. In this case the direction of the electric field rotates in time while its magnitude remains constant. There are two types of circular polarization—right-hand circular polarization (RHCP) and left-hand circular polarization (LHCP). With RHCP, the E field rotates in the direction of the curled fingers of the right hand when the right thumb points in the direction of wave propagation (i.e., the direction of $\mathbf{k}$). With LHCP, E rotates in the direction indicated by the curled fingers of the left hand when the left thumb points along $\mathbf{k}$. For outward propagating waves with $\mathbf{k} = k\,\mathbf{a}_r$, the polarization of a RHCP wave is designated by the polarization vector

$$\mathbf{h} = \sqrt{\frac{1}{2}}(\mathbf{a}_\theta - j\,\mathbf{a}_\phi) \tag{3.94}$$

For a LHCP wave,

$$\mathbf{h} = \sqrt{\frac{1}{2}}(\mathbf{a}_\theta + j\,\mathbf{a}_\phi) \tag{3.95}$$

If we designate the polarization of the incoming transmitted wave as $\mathbf{h}_t$ and the polarization that the receiving antenna is configured to accept as $\mathbf{h}_r$, then the polarization mismatch factor by which the ratio P_i/P_t will be reduced is defined by

$$K_p = |\,\mathbf{h}_t \bullet \mathbf{h}_r^*\,|^2 \tag{3.96}$$

The Friis transmission formula would then be modified to read

$$P_i/P_t = (16\,\pi^2)^{-1}\,(\lambda/d)^2\,G_t\,G_r\,K_p \tag{3.97}$$

As an example, consider the case where the incoming wave is RHCP and the receiving antenna is designed to receive RHCP waves. Then, the polarization mismatch factor is

$$K_p = \tfrac{1}{2}\,(\mathbf{a}_\theta - j\,\mathbf{a}_\phi) \bullet (\mathbf{a}_\theta + j\,\mathbf{a}_\phi)$$

$$= \tfrac{1}{2}\,(1 + 1)$$

$$= 1$$

On the other hand, if the incoming wave were RHCP while the receiving antenna was designed to accept LHCP waves,

$$K_p = \tfrac{1}{2}\,(\mathbf{a}_\theta - j\,\mathbf{a}_\phi)\bullet(\mathbf{a}_\theta - j\,\mathbf{a}_\phi)$$

$$= \tfrac{1}{2}\,(1 - 1)$$

$$= 0$$

Similarly, if the incoming wave were linearly polarized with $\mathbf{h}_t = \mathbf{a}_\theta$ while the receiving antenna were polarized with $\mathbf{h}_r = \mathbf{a}\hat{\phi}$, we would again have $K_p = 0$.

If the incoming wave were circularly polarized with for example $\mathbf{h}_t = \sqrt{\tfrac{1}{2}}(a_\theta + j\ a\hat{\phi})$, and the receiving antenna were linearly polarized with say $\mathbf{h}_r = \mathbf{a}\hat{\theta}$, then

$$K_p = \left|\sqrt{\frac{1}{2}}(a_\theta + j\,a_\phi)\cdot a_\phi\right|^2$$

$$= 1/2$$

Thus, we see that polarization matching is of crucial importance to receiving an acceptable signal strength. In satellite communications (SATCOM) where a linearly polarized wave might have its direction of polarization rotated by an unpredictable amount, circularly polarized waves are used.

3.3.4 A Brief Treatment of Aperture Antennas

There is an important class of antennas in which the source of the radiation field is the electromagnetic field pattern in the antenna aperture. Examples of such antennas as shown in Figure 3.10 are a horn antenna at the end of a waveguide, and a parabolic reflector illuminated by the end of a waveguide located at the focal point. One might expect that when such antennas are used for receiving a signal that their effective area would be approximately equal to the physical area of the aperture, A_{ap}, multiplied by some factor smaller than but on the order of one, to account for the mismatch between the configuration of the aperture fields and the almost uniform field in the incoming wave; that is,

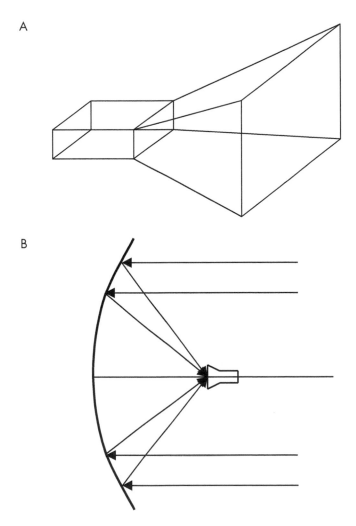

FIGURE 3.10 (a) Horn antenna. (b) Parabolic reflector antenna. Arrows show direction of rays when antenna is receiving.

$$A_e \lesssim A_{ap} \qquad\qquad (3.98)$$

This relationship combined with Equation (3.90) allows one to quickly estimate the gain of any aperture antenna if the signal frequency is known. For example, a radar antenna with a radius of 30 cm and operating at 10 GHz would have a gain $G \lesssim 400\,\pi^2$ or approximately 36 dB. Thus, it is clear that even moderately sized aperture antennas can have a large gain.

A typical radiation pattern of an aperture antenna is shown in Figure 3.11.

3D spherical plot of
directivity (dimensionless)

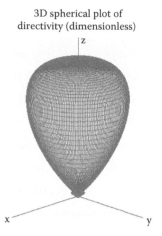

FIGURE 3.11 Typical radiation pattern of an aperture antenna. Aperture in the *x-y* plane and the center of the aperture at the origin.

The axes are aligned so that the aperture is located in the *x-y* plane with its center at the origin; the far field pattern is maximum along the *z*-axis. Formal analysis that allows for prediction of the far fields from the source fields in the aperture is available in the literature (e.g., Sturtzman and Thiele, 1981). For our present purposes, we will use the result that for the case of large aperture dimensions compared with the wavelength, the maximum power density along the *z*-axis is given by

$$S_{max} = (2\,Z_o\,r^2\,\lambda^2)^{-1}\,|\iint_{S_a} E_a(x',y')\,dx'\,dy'\,|^2 \qquad (3.99)$$

From the form of Equation (3.99) we see that the power density in the propagating wave falls off as $1/r^2$ as expected in the far field of an antenna. The integral is over the electric field in the aperture E_a and covers the entire aperture area, S_a. Thus, we see that each point on the wavefront in the aperture contributes toward the far field on the *z*-axis. This is in agreement with Huygens' principle, which states that each point on a wavefront is a secondary source of a spherical wave.

The total radiated power, P_r, passing through the aperture is

$$P_r = (2\,Z_o)^{-1}\iint_{S_a}|E_a(x',y')|^2\,dx'\,dy' \qquad (3.100)$$

Then, using Equation (3.99) and Equation (3.100) we can define the directivity of an aperture antenna as

$$D_a = 4\pi \ r^2 \ S_{max}/P_r = (4\pi/\lambda^2) \ \frac{\left| \iint_{S_a} E_a(x',y') \ dx' \ dy' \right|^2}{\iint_{S_a} |E_a(x',y')|^2 \ dx' \ dy'} \qquad (3.101)$$

Then from Equation (3.90), the effective area of an aperture antenna is

$$A_e = \eta \ \frac{\left| \iint_{S_a} E_a(x',y') \ dx' \ dy' \right|^2}{\iint_{S_a} |E_a(x',y')|^2 \ dx' \ dy'} \qquad (3.102)$$

Clearly, if the aperture fields were uniform, Equation (3.102) would give $A_e = \eta \ A_{ap}$.

Let us consider the case of a waveguide horn antenna with aperture dimensions a and b, where a is the wide dimension and is assumed to be aligned with the x-axis. If the horn taper is sufficiently gradual, we may assume that the aperture field has the same form as the dominant TE_{10} mode in a rectangular waveguide; that is,

$$E_a = E_o \sin(\pi x/a) \ \mathbf{a}_y \qquad (3.103)$$

where E_o is the amplitude of the electric field at the center of the aperture.

Substituting Equation (3.103) into Eqution (3.102) gives

$$A_e = \eta \ \frac{b^2 \ |E_o|^2 \left[\int_0^a \sin(\pi x'/a) \ dx' \right]^2}{b \ |E_o|^2 \int_0^a \sin^2(\pi x'/a) \ dx'}$$

$$= 0.81 \, \eta \, ab$$

$$= 0.81 \, \eta \, A_{ap}$$

$$= \eta \, \eta_{ap} A_{ap} \tag{3.104}$$

The quantity η_{ap} is called the aperture efficiency. Clearly for a rectangular horn antenna, the aperture efficiency $\eta_{ap} = 81\%$. In general, aperture efficiency in accord with Equations (3.89) and (3.104) is defined by

$$\eta_{ap} = (A_{ap})^{-1} \, \frac{\left| \iint\limits_{S_a} E_a(x',y') \, dx' \, dy' \right|^2}{\iint\limits_{S_a} \left| E_a(x',y') \right|^2 \, dx' \, dy'} \tag{3.105}$$

In addition to rectangular horn antennas, for other aperture antennas, calculations similar to that made in Equation (3.104) will be appropriate, but the spatial variation of the aperture fields and the resulting value of aperture efficiency will in general be different. For example, in the case of the parabolic reflector antenna, the aperture efficiency is 55%. For the case of aperture antennas with large dimensions compared to the wavelength, the efficiency based on Ohmic losses, η, will be close to 100%, and aperture efficiency will be the more important parameter to evaluate.

As an example, let us consider a parabolic reflector antenna with a diameter of 50 cm operating at a frequency of 12.4 GHz ($\lambda = 2.42$ cm) with Ohmic efficiency near 100%. The physical antenna area is $A_{ap} = 0.196$ m^2 and the effective area is 55% of that, or $A_e = 0.108$ m^2. The corresponding value of antenna gain is $G = 4\pi \, A_e / \lambda^2 = 4200$. Because Ohmic losses are negligible, this is also approximately the value of the antenna directivity D_a. When the antenna is used in transmission it will produce a narrow radiation beam with circular cross section (i.e., $BW_\phi = BW_\theta$). The beamwidth may be estimated using Equation (1.11) to obtain

$$BW_\phi(\text{deg}) = BW_\theta \, (\text{deg})$$

$$\approx [41000/D_a]^{1/2}$$

$$\approx [41000/4200]^{1/2}$$

$$= 3°$$

3.4 Thin-Wire Dipole Antennas

Aperture antennas are widely used in such applications as SATCOM, radar, and space research, but they are not generally used in cell phone communications because of their tightly focused radiation beam. In cell phone systems, the direction from the base station to the mobile units is largely confined near the horizontal plane but is otherwise unknown. Therefore, what is needed is an antenna with a radiation pattern that is omnidirectional in the horizontal plane. The radiation pattern of a short dipole antenna such as that discussed in Section 3.2.6 is of the desired omnidirectional shape. However, as we saw, the radiation resistance of a short dipole might be a small fraction of an Ohm. This presents two problems. First, it would be very difficult to match the impedance of the antenna to the impedance of a feed cable; cable characteristic impedances are on the order of 100 Ohms. Second, the Ohmic efficiency of the antenna is likely to be significantly smaller than 100%. Also, short antennas have capacitive reactance that must be compensated by adding series inductive loops as pictured in Figure 3.12.

To overcome these problems while still retaining the omnidirectional radiation pattern, antennas closely related to the "half-wave" dipole, which is approximately one half wavelength long, have become the basic antenna type in cell phone communication systems. Base stations employ arrays of half-wave dipole antennas while the mobile transceivers effectively use a quarter-wave dipole above an image plane.

The reactance of a dipole antenna as a function of its length is shown in Figure 3.13. Antenna wire thickness is the parameter that changes on the various curves. The most important feature to note in this figure is that all the curves for the various wire thicknesses pass through the line of zero reactance for antenna length just slightly smaller than one half wavelength ($\ell = 0.48\ \lambda$). A dipole antenna of this length is known as a resonant dipole or "half-wave" dipole. Shorter antennas have capacitive reactance, while longer antennas have inductive reactance as may be seen from Figure 3.13.

FIGURE 3.12 Short antenna with capacitive reactance compensated by inductive loops.

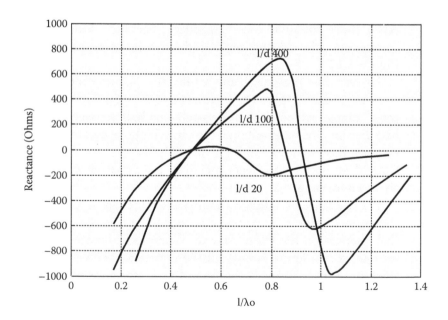

FIGURE 3.13 Input reactance of a dipole antenna of length *l* and diameter *d*. (from R.E. Collin, *Antennas and Radiowave Propagation*, McGraw-Hill Book Company, New York, 1985. Reproduced with permission of the McGraw-Hill Companies.).

3.4.1 General Analysis of Thin-Wire Dipole Antennas

We will now analyze the thin-wire dipole antenna without the restriction of antenna length being much shorter than the wavelength that was used in Section 3.2.6. This will allow radiation emanating from different positions along the dipole to differ in phase when arriving at the observation point in the far field. The antenna is shown in Figure 3.14 with its central feedpoint located at the origin of a spherical coordinate system (r, θ, ϕ). The antenna wire is aligned along the z-axis. Each incremental length along the current wire is specified by its position z' and its coordinates with respect to the observation point (r', θ') Current along the wire vanishes at the two ends of the wire so that the current variation along the wire will be as given previously by Equation (3.53), which we repeat here for convenience:

$$I(z) = I_i \frac{\sin\left[k(h - |z|)\right]}{\sin(k\,h)}, -h<z<h \qquad (3.106)$$

The current distribution along the dipole antenna is plotted in Figure 3.15 for the following three cases of antenna length $2h$: (a) $2h < \lambda/2$: (b) $2h = \lambda/2$; and (c) $2h > \lambda/2$.

In the far field, the contribution to the radiation electromagnetic field from a differential current element $I(z)\,dz$ may be obtained from our previous analysis of the Hertzian dipole as

$$dE_\theta = Z_o\,dH_\phi = \frac{j\,Z_o\,k^2\,I(z)\,dz\,e^{-jkr'}\sin\theta'}{4\pi\,k\,r'} \qquad (3.107)$$

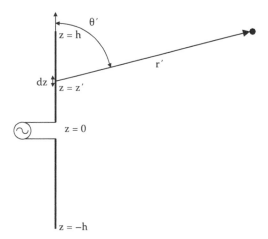

FIGURE 3.14 Dipole antenna geometry.

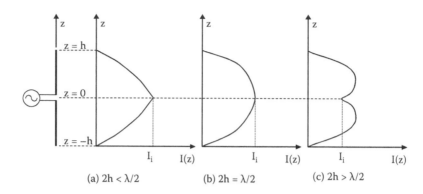

FIGURE 3.15 Current distributions along dipole antennas of various lengths.

Now, r' may be related to the spherical coordinates by considering the triangle shown in Figure 3.16 and applying the well-known trigonometric relationship between the sides and angles of any triangle to get

$$r' = (r^2 + z^2 - 2rz\cos\theta)^{1/2} \qquad (3.108)$$

In accord with the far field conditions described by Equation (3.40), the distance from the antenna to the observation point must greatly exceed the length of the antenna, so we assume that for all values of z where the current is nonzero, $r' \gg z$, so that Equation (3.108) yields the approximate equality

$$r' \cong r - z\cos\theta, \quad -h < z < h \qquad (3.109)$$

Then, Equation (3.106) may be written as

$$dE_\theta = Z_o\, dH_\phi = \frac{j\, Z_o\, k^2\, I(z)\, dz\, e^{-jkr+jkz\cos\theta}\sin\theta}{4\pi\, k\, r} \qquad (3.110)$$

In writing Equation (3.110) we have assumed that the small difference between r' and r as reflected by Equation (3.109), and the corresponding small difference between θ' and θ, have a negligible effect on the far field amplitude, but might have an important effect on phase. When field contributions from different parts of the antenna are summed, phase difference could result either in reinforcement or in cancellation.

To find the total field from all the current elements along the antenna, we substitute the expression for $I(z)$ from Equation (3.106) into Equation (3.110) and integrate along z from $-h$ to h:

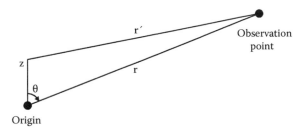

FIGURE 3.16   Relationship between r', r, z, and θ.

$$E_\theta = Z_o \, H_\phi = \frac{j \, Z_o \, k^2 \, I_i \, e^{-j \, k \, r} \sin\theta}{4\pi \, k \, r \, \sin kh} \int_{-h}^{h} \sin k(h - |z|) e^{jkz \cos\theta} \, dz$$

(3.111)

The integral in Equation (3.111) can be performed analytically with the result

$$E_\theta = Z_o \, H_\phi = \frac{j \, Z_o \, I_i \, e^{-j \, k \, r} f(\theta)}{2\pi \, r \, \sin kh}$$

(3.112)

where

$$f(\theta) = [\cos(kh \cos\theta) - \cos kh]/\sin\theta$$ (3.113)

$f(\theta)$ is called the pattern function of a linear dipole.

3.4.2 The Half-Wave Dipole

When the dipole antenna length $\ell = 2h = \lambda/2$, Equations (3.112) and (3.113) are somewhat simplified. Note that in this "half-wave" dipole case, $kh = (2\pi/\lambda)(\lambda/4) = \pi/2$, so that Equation (3.112) becomes

$$E_\theta = Z_o \, H_\phi = j \, Z_o \, I_i \, f(\theta) \, e^{-jkr}/(2\pi \, r)$$ (3.114)

and Equation (3.113) becomes

$$f(\theta) = f_h(\theta) = \cos[(\pi/2)\cos\theta]/\sin\theta$$ (3.115)

Thus, the pattern function of a half-wave dipole antenna $f(\theta) = 1$ when $\theta = \pi/2$ (i.e., in the horizontal plane). When $\theta = 0$ or when

$\theta = \pi$, the pattern function can be evaluated using l'Hospital's rule to determine that $f(\theta) = 0$. Thus, the pattern function is omnidirectional and maximum in the horizontal plane as desired for cellular communication systems.

If we substitute Equation (3.115) into Equation (3.114), we get the following expression for the far fields of the half-wave dipole antenna:

$$E_\theta = Z_o H_\phi = j Z_o I_i [e^{-jkr}/(2\pi r)] [f_h(\theta)/\sin \theta] \qquad (3.116)$$

Using this equation, the following expression is obtained for the time-average radiated power density:

$$S = \frac{1}{2} E_\theta H_\phi^* = \frac{1}{2} |E_\theta|^2/Z_o = |I_i|^2 Z_o f_h(\theta)^2/(8 \pi^2 r^2) \qquad (3.117)$$

The total radiated power is then given by applying Equation (3.46):

$$P_r = \int_0^{2\pi} \int_0^{\pi} S r^2 \sin\theta \, d\theta \, d\phi \qquad (3.118)$$

The integral in Equation (3.118), with S as specified in Equation (3.117), has been evaluated numerically with the result

$$P_r = 36.54 |I_i|^2, \text{ mks units} \qquad (3.119)$$

The radiation resistance of the half-wave dipole is then readily evaluated using Equation (3.119) as

$$R_r = 2 P_r/|I_i|^2 = 73.08 \text{ Ohms}$$

The radiation resistance for the half-wave dipole is seen to be much larger than the radiation resistance of an electrically short dipole. In fact, a radiation resistance of $R_r = 73$ Ohms is readily matched to the characteristic impedance of a transmission line feeding the half-wave dipole. This advantage, together with the resonant cancellation of reactive impedances, recommends the half-wave dipole antenna for application in many systems.

The directivity of the half-wave dipole is obtained by noting that the maximum value of radiated power density is obtained from Equation (3.117) by setting $\theta = \pi/2$ so that the quantity in square brackets equals 1. Then

$$D = 4\pi \, r^2 \, S_{max}/P_r = 1.64$$

The corresponding value of beamwidth may be obtained from Equation (1.13) as $BW\hat{\theta} = 79°$. This value of directivity is only marginally larger than the directivity of a short dipole, and the beamwidth is only marginally smaller ($79°$ versus $90°$).

To achieve larger values of directivity (i.e., a more tightly focused radiation beam), arrays of dipole antennas are used on cell phone base station towers; antenna arrays will be described in Chapter 4. The handheld cell phones effectively employ a quarter-wave monopole antenna over a ground plane, which has strong similarity to the half-wave dipole; this type of antenna is in the category of image antennas that are also discussed in Chapter 4.

Problems

3.1. Calculate the distance that a plane wave of frequency 1 GHz must travel in free space in order for its phase to be retarded by $90°$. Repeat the calculation for propagation in a lossless dielectric with relative permittivity of nine.

3.2. Calculate the average power density and the magnetic field strength in a plane wave propagating in free space with peak electric field strength of 1 millivolt/meter. Repeat the calculation for propagation in a lossless dielectric with relative permittivity of four.

3.3. By what factor is a 1 GHz plane wave attenuated after traveling for 10 m through a dielectric medium with relative permittivity of nine, and loss tangent of 10^{-4}. What is the conductivity?

3.4. Prove by substitution that the solution to the ordinary differential equation

$$\frac{1}{r^2} \frac{\partial}{\partial r} (r^2 \frac{\partial A}{\partial r}) + k^2 A = 0$$

is

$$A = C_1 e^{-jkr}/r + C_2 e^{jkr}/r, \ (r \neq 0)$$

3.5. Show that

$$\int_0^{r_0} Ak^2 r^2 dr \to 0 \text{ as } \to 0$$

where

$$A = C_1 e^{-jkr} / r, (r \neq 0)$$

3.6. An aluminum dipole antenna is 10 cm long and 1 mm thick. It is excited by a radio frequency (RF) current with a feedpoint rms value of 1 ampere at a frequency of 50 MHz. Assuming a triangular pattern of current along the antenna length, calculate the following:
   a. Antenna efficiency
   b. Antenna gain
   c. Radiated power
   d. Transmitter power

3.7. How far from a dipole antenna, with length equal to one half of a wavelength (half-wave dipole), does one have to be for far field approximations to be valid? Express your answer in wavelengths.

3.8. Calculate the maximum power density 1 km away from a half-wave dipole antenna that is radiating 100 milliwatts. Repeat the calculation for a dipole antenna that is much shorter than a wavelength. (The gain of a half-wave dipole antenna is 2.15 dBi.)

3.9. Calculate the voltage standing wave ratio (VSWR) for a small, reactance-compensated, dipole antenna with length = 0.1 λ fed from a 75 Ohm line. Calculate the fraction of power incident on the antenna that will be reflected back to the transmitter.

[When transmission line impedance, Zo, is not equal to the impedance of the antenna at the end of the line, Zr, there will be reflection of a portion of the wave energy, and a standing wave will be set up on the transmission line. At the antenna, the ratio of the electric field in the reflected wave to the electric field in the incident wave is known as the reflection coefficient, given by $\rho = (Z_r - Z_0)/(Z_r + Z_0)$. On the transmission line, the standing wave pattern will have maxima and minima in the electric field. The ratio of the maximum electric field to the minimum electric field is called the VSWR and is given by VSWR = $(1 + |\rho|)/(1 - |\rho|)$.]

3.10. Repeat Problems 3.9 for a half-wave antenna whose radiation impedance is 73 Ohms.

3.11. An antenna with a gain of 3 dBi is in a region where incident power density is 1 microwatt per square meter. Calculate the voltage across the matched input resistance of the receiver fed by the antenna. The frequency is 900 MHz and the radiation resistance is 9 Ohms.

3.12. A 870 MHz base station antenna array has vertical dimension of 2.5 m and horizontal dimensions of 0.4 m. What is the maximum possible gain?

3.13. What is the polarization mismatch factor for two linearly polarized antennas whose polarization vectors are mismatched by an angle of 35°?

3.14. Show how two linearly polarized antennas and a phase shifter can be arranged to receive an incoming wave with circular polarization without signal attenuation due to polarization mismatch.

3.15. A 100 Watt transmitter feeds an antenna with a gain of 50 dBi. Calculate the maximum power that could be received at a geosynchronous satellite at an altitude of 36,000 km whose antenna has a gain of 30 dB. The wavelength is 1 cm, and the polarization mismatch factor is −3 dB.

3.16. Prove that

$$\int_{-h}^{h} \sin\,[k(h - |z|)]e^{jkz\,\cos\theta}\;dz = 2f(\theta)/[k\,\sin\theta]$$

3.17. Use l'Hospital's rule to find the limiting value of the pattern function for a half-wave dipole as $\theta \Rightarrow 0$.

3.18. A loop antenna consists of 15 turns of 1 mm thick copper wire wound on a ferrite core with a radius of 1 cm and with relative permeability of 40. The frequency is 100 MHz. Neglecting losses in the ferrite, calculate the antenna efficiency.

Bibliography

1. D.K. Cheng, *Field and Wave Electromagnetics*. 2nd ed. (Addison-Wesley, Reading, Massachusetts, 1989), 602–638.
2. W.L. Sturtzman and Gary A. Thiele, *Antenna Theory and Design* (John Wiley & Sons, New York, 1981), 79–94.
3. R.E. Collin, *Antennas and Radiowave Propagation* (McGraw-Hill, New York, 1985), 32–103.
4. J.D. Kraus, *Antennas* (McGraw-Hill, New York, 1950).
5. S.A. Schelkunoff, *Advanced Antenna Theory* (John Wiley & Sons, New York, 1952).

CHAPTER 4

Antenna Arrays

Antenna arrays are groups of *similar* antennas arranged in various spatial configurations with relative amplitude and phase between the exciting currents or fields driving adjacent antennas chosen to give a desired radiation pattern. We will concentrate in this chapter on arrays of half-wave dipole antennas; however, many of the features of array radiation that will be uncovered are universally applicable to arrays of any type of antenna as will become clear. The chapter will conclude with a discussion of microstrip patch antennas that can be fabricated by inexpensive lithographic techniques and are replacing aperture antennas in many applications; microstrip patch antennas can be analyzed as a two-dimensional array of dipole antennas.

4.1 Omnidirectional Radiation Pattern in the Horizontal Plane with Vertical Focusing

The antennas used in cell phone base stations are typically arrays of half-wave dipoles often arranged to maintain an omnidirectional radiation pattern in the horizontal plane (i.e., the radiation pattern is not a function of ϕ), but with stronger focusing in the direction of the polar angle θ than can be achieved with a single dipole. Such arrays will be described below.

4.1.1 Arrays of Half-Wave Dipoles

An array of antennas is pictured in Figure 4.1. The position vector from the nth antenna is denoted by the vector $\mathbf{r}_n$ from the origin to the antenna's location. The vector from the nth antenna to the observation point P in the far field is designated by $\rho(n) = \mathbf{r} - \mathbf{r}_n$, and its magnitude is $\rho(n) = |\mathbf{r} - \mathbf{r}_n|$.

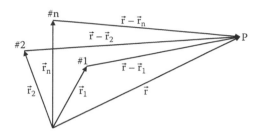

FIGURE 4.1 Vectors for an array of n antennas.

Let us specifically consider an array of linear, half-wave, dipole antennas with wires oriented parallel to the z-axis. Recall that for a dipole located at the origin, the far fields produced are given by Equation (3.101) as

$$E_\theta = Z_o \, H_\phi = j \, Z_o \, I_i \, [e^{-jkr}/(2\pi r)] \, f(\theta) \qquad (4.1)$$

Similarly, for the nth antenna displaced from the origin by $\mathbf{r}_n$ and with input current I_n, the far field produced will be

$$E_{\theta n} = Z_o \, H_{\phi n} = j \, Z_o \, I_n \, [e^{-jk\,\rho(n)}/(2\pi \, \rho(n))] \, f(\theta_n) \qquad (4.2)$$

We assume that the displacement of any array element from the origin is small compared with the distance from the origin to the far field observation point. This assumption implies that $\rho(n)$ in the denominator in Equation (4.2) may be replaced by r. Also, θ_n may be replaced by θ. Only the phase differences of the field contributions from the different array antennas are retained. Then, summing over the contributions to the far field from all the antennas in the array gives the total fields produced by the array as

$$E_\theta = Z_o H_\phi = [jZ_o/(2\pi r)] f(\theta) \sum_{n=0}^{N-1} I_n e^{-jk\rho(n)} \qquad (4.3)$$

4.1.2 Colinear Arrays

For a radiation pattern that is omnidirectional in the horizontal plane, but tightly focused vertically, colinear arrays of half-wave dipoles may be employed. Simply put, in a colinear array, the dipole antennas are displaced along the z-axis and all the antenna wires are oriented so that currents flow in the z-direction.

The colinear array geometry is shown in Figure 4.2. Using the trigonometric relationship between the sides and angles of a triangle, the magnitude of the position vector from the nth antenna to the far field observation point may be expressed as

$$\rho(n) = [z_n^2 + r^2 - 2\,z_n\,r\,\cos\theta]^{1/2}$$

$$\approx [r^2 - 2\,z_n\,r\,\cos\theta]^{1/2}$$

$$\approx r\,(1 - z_n\,\cos\theta) \tag{4.4}$$

where we have assumed that $r \gg z_n$.

Then, substituting from Equation (4.4) into Equation (4.3) gives

$$E_\theta = Z_o H_\phi = j\,Zo\,f(\theta)[e^{-jkr}/(2\pi\,r)] \sum_{n=0}^{N-1} I_n \exp(jkz_n\cos\theta) \tag{4.5}$$

For the special case where all the antennas in the array are excited by currents having identical magnitude and phase (i.e., $I_n = I_i$), Equation (4.5) becomes

$$E_\theta = Z_o\,H_\phi$$

$$= j\,Z_o\,I_i f(\theta)\,[e^{-jkr}/(2\pi\,r)] \sum_{n=0}^{N-1} \exp(jkz_n\cos\theta) \tag{4.6}$$

$$= E_{\theta 1}\,A(\theta)$$

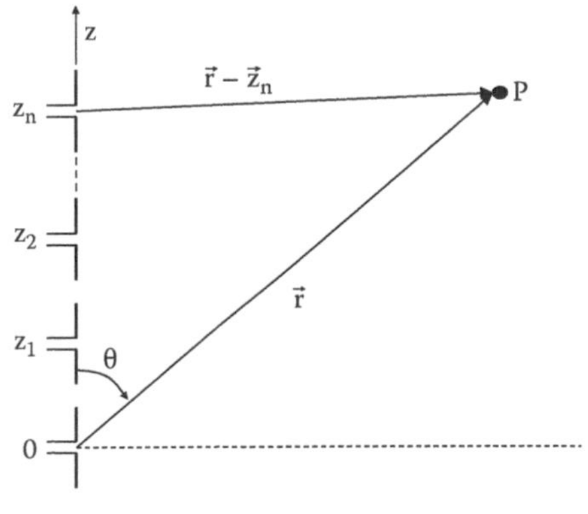

FIGURE 4.2 Collinear array geometry.

where $E_{\theta 1}$ is the far field produced by a single antenna as previously given by Equation (4.1) and $A(\theta) = \sum_{n=0}^{N-1} \exp(jkz_n \cos\theta)$ is called the array pattern function.

The form of Equation (4.6) is quite general for an array of antennas with identical excitation (i.e., the far field is given by the product of the far field produced by a single antenna and an array pattern function that depends on the positions of the antennas in the array).

4.1.3 Colinear Arrays with Equal Incremental Phase Advance

Another very important result for a colinear array is obtained if the antennas are equally spaced, and here is a linear phase advance with increment α between adjacent antennas—that is, $z_n = n \, d_V$ where d_V is the spacing between adjacent antennas, and $I_n = I_i \exp(jn\alpha)$.

In that case, the far field becomes

$$E_\theta = Z_o \, H_\phi$$

$$= j \, Z_o I_i f(\theta)[e^{-jkr}/(2\pi r)] \sum_{n=0}^{N-1} \exp(jn\chi) \qquad (4.7a)$$

$$= E_{\theta 1} A(\theta)$$

where

$$\chi = (kd_V \cos\theta + \alpha) \qquad (4.7b)$$

As in Equation (4.6), Equation (4.7) also gives the far field as a product of the far field produced by a single antenna and an array pattern function, but the array pattern function is now

$$A(\theta) = \sum_{n=0}^{N-1} \exp(jn\chi) = 1 + e^{j\chi} + \dots e^{j(N-1)\chi} \qquad (4.8)$$

The series in Equation (4.8) may be easily summed. First multiply both sides of this equation by $e^{j\chi}$ to get

$$A(\theta) \, e^{j\chi} = e^{j\chi} + e^{j2\chi} + \dots + e^{jN\chi} \qquad (4.9)$$

Next subtract Equation (4.9) from Equation (4.8) to get

$$A(\theta) \, [1 - e^{j\chi}] = 1 - e^{jN\chi}$$

or

$$A(\theta) = \frac{1 - e^{jN\chi}}{1 - e^{j\chi}} = \frac{e^{jN\chi/2} \left(e^{-jN\chi/2} - e^{jN\chi/2} \right)}{e^{j\chi/2} \left(e^{-j\chi/2} - e^{j\chi/2} \right)}$$

$$= e^{j(N-1)\chi/2} \frac{\sin(N\chi / 2)}{\sin(\chi / 2)} \tag{4.10}$$

The phase factor $e^{j(N-1)\chi/2}$ is unimportant unless the signal from the colinear array is going to be combined with some other signal. If not, the amplitude of the array factor is what matters: the expression for the amplitude is

$$|A(\theta)| = \frac{\sin(N\chi / 2)}{\sin(\chi / 2)} \tag{4.11}$$

Let us now consider the electric field strength produced by the colinear array in the horizontal plane (i.e., when $\theta = \pi/2$) and with zero phase advance between adjacent antennas. Clearly, from Equation (4.7b), this corresponds to $\chi = 0$. The magnitude of the array pattern function may then be found from Equation (4.11) by taking the limit as $\chi \to 0$:

$$\lim_{\chi \to 0} |A(\theta)| = \left| \lim_{\chi \to 0} \frac{\sin(N\chi / 2)}{\sin(\chi / 2)} \right| \approx N \tag{4.12}$$

Then, it is easily seen from Equation (4.7a) that the magnitude of the far field becomes

$$|E_\theta| = Z_o \, | H_\phi | = | E_{\theta 1} | \, | A | = | E_{\theta 1} | \, N \tag{4.13}$$

Equation (4.13) states that the electric field produced by N in-phase antennas is N times the field produced by a single antenna. Correspondingly, because power density in the radiation field is proportional to electric field magnitude squared, the power density is enhanced by a factor of N^2 over the power density produced by a single radiator. This may be compared with N incoherent sources, where the combined power density is simply N times the power of a single source. Thus, Equation (4.13) expresses an important fundamental property of all coherent radiating sources. For example, a laser utilizes radiation from excited atoms or molecules and is said to be lasing (i.e., producing a high-power density, tightly focused beam) when its atoms or molecules are radiating in phase.

The maximum power density produced by the array of N coherent radiators described by Equation (4.13) is

$$S_{max} = |E_{\theta, max}|^2/(2\,Z_o)$$

$$= N^2\,|E_{\theta 1 max}|^2/(2\,Z_o)$$

$$= N^2\,S_{1\,max} \tag{4.14}$$

On the other hand, the total power radiated by the array, P_{rA}, is simply the power radiated by a single antenna, P_{r1}, times the number of antennas; that is,

$$P_{rA} = N P_{r1} \tag{4.15}$$

Then, total directivity of the array is found by using Equations (4.14) and (4.15):

$$D = 4\pi\,r^2\,S_{max}/P_{rA}$$

$$= 4\pi\,r^2\,S_{1\,max}\,N^2/(P_{r1}N)$$

$$= D_1\,N \tag{4.16}$$

where $D_1 = 1.64$ is the directivity of each half-wave dipole antenna. Thus, we see that the array directivity is 1.64 N.

As an example, consider a colinear array of 10 half-wave dipoles. The total directivity of the array will be 16.4. To estimate the beamwidth we will use Equation (1.13), but assume that the directivity is sufficiently large and the beamwidth sufficiently small that $BW_\theta(\deg) \approx 0.0027\,[BW_\theta(\deg)]^2$. In that case, Equation (1.13) becomes $BW_\theta(\deg) \approx 102°/D$. In the present example with $D = D_A = 16.4$, $BW_\theta(\deg) \approx 6.2°$.

Thus, the radiation pattern is omnidirectional (i.e., not a function of ϕ), but is sharply focused in the θ direction around θ = π/2.

4.1.4 Elevation Control with a Phased Colinear Antenna Array

As presented above, an array of colinear dipole antennas with identical excitation of each antenna can produce a narrow omnidirectional radiation pattern. Such a pattern will correspond to radiation that is uniform and maximum in the horizontal plane passing through the array. However, in a cell phone system, the base station antenna array is usually mounted at the top of a tall tower, while the mobile units with which the base station needs to communicate are on the ground. Thus, it is of interest to have a means for tilting the base station radiation pattern downward. This can be accomplished by instituting an incremental phase advance between adjacent antennas as described in Section 4.1.3.

Consider the array pattern function in Equation (4.11):

$$|A(\theta)| = \frac{\sin(N\chi / 2)}{\sin(\chi / 2)}$$

which will be maximum when θ = θ_{max} as defined by

$$\chi = k \, d_V \cos \theta_{max} + \alpha = 0 \tag{4.17}$$

Consider the example of a colinear array of 10 half-wave dipoles with center-to-center spacing between adjacent antennas of d_V = 0.75 λ (i.e., kd_V = 1.5 π). Suppose a downward tilt in the radiation pattern of 4° is desired. Then, one wants a maximum in the array pattern when θ = θ_{max} = 94° = 1.64 radians. Equation (4.17) can then be used to calculate the required incremental phase advance:

$$\alpha = -k \, d_V \cos (1.64)$$

$$= -1.5 \, \pi \times (-0.0698)$$

$$= 0.329 \text{ radians}$$

If this antenna array were mounted on top of a 50 m high tower (i.e., h_b= 50 m), the maximum in the radiation pattern would strike the ground at a distance from the base of the tower equal to

$$\rho_{max} = h_b/\tan(\theta_{max} - 90°) \tag{4.18}$$

For the present example, Equation (4.18) gives ρ_{max} = 50 m/tan 4° = 715 m. With a beamwidth of ~6°, an annular-shaped area of considerable extent on the ground around the base of the antenna tower would be illuminated with sufficient power to establish a successful communications link between the base station and mobile transceivers (cell phones). However, for regions near the tower base, nulls will be encountered in the radiation pattern.

According to the form of the array pattern function, as one approaches the tower, the first null will occur when the polar angle $\theta = \theta_{FN}$ such that

$$N\chi/2 = N(k\,d_V\cos\theta_{FN} + \alpha)/2 = -\pi$$

or

$$\theta_{FN} = \cos^{-1}\{[-(2\pi/N) - \alpha]/(k\,d_V)\} \tag{4.19}$$

For the parameters of the example in the preceding paragraph (viz., N = 10; α = 0.329 radians; k d_V = 1.5 π), Equation (4.19) gives $\theta_{FN} = \cos^{-1}(-0.203) = 101.7°$. Thus, the region around the base of the tower, in which nulls occur, extends out to a radius

$$\rho_{FN} = h_b/\tan(\theta_{FN} - 90°) \tag{4.20}$$

which for the present example gives ρ_{FN} = 240 m.

Thus, we have achieved illumination on the ground with a power density maximum at a radius of 715 m from the antenna tower but with nulls in the radiation occurring out to a radius of 240 m from the tower. The close-in region could be illuminated by auxilliary downward-facing antennas. The geometry discussed in this section is illustrated in Figure 4.3.

4.2 Antennas Displaced in the Horizontal Plane

The preceding sections of this chapter concerned antenna arrays that produced radiation patterns that have omnidirectionality in the horizontal plane. That property is not always desirable even in cell phone systems. For example, the center of a cell may not be accessible, and the base station may have to be located at the edge of the cell rather than at the center. Another case is when sectorization of a cell is used to minimize call interference as discussed in Section 1.3.1.

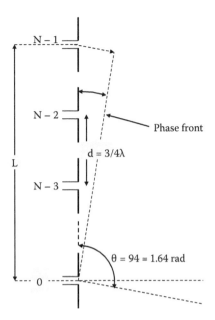

FIGURE 4.3 Collinear array with phase advance producing downward tilt in a radiation pattern.

4.2.1 Radiation Pattern of Two Horizontally Displaced Dipoles

The simple example of two half-wave dipole antennas displaced along the *x*-axis by a distance d_H is shown in Figure 4.4. The position vector of antenna #1 is

$$\mathbf{r}_1 = -(d_H/2)\, \mathbf{a}_x \qquad (4.21)$$

and the position vector of antenna #2 is

$$\mathbf{r}_2 = (d_H/2)\, \mathbf{a}_x \qquad (4.22)$$

We can insert these values into Equation (4.3), which is the general expression for the far field from an array of half-wave dipoles, to get

$$E_\theta = [\mathrm{j}\, Z_o/(2\pi\, \mathrm{r})]\, \mathrm{f}(\theta) \sum_{n=0}^{N-1} I_n \mathrm{e}^{-\mathrm{j}k\,|\mathbf{r}-\mathbf{r}_n|}$$

$$= [\mathrm{j}\, Z_o/(2\pi\, \mathrm{r})]\, \mathrm{f}(\theta)\, \{I_1\, \mathrm{e}^{-\mathrm{j}k\,|\mathbf{r}-\mathbf{r}_1|} + I_2\, \mathrm{e}^{-\mathrm{j}k\,|\mathbf{r}-\mathbf{r}_2|}\}$$

$$= [\mathrm{j}\, Z_o/(2\pi\, \mathrm{r})]\, \mathrm{f}(\theta)\, \{I_1\, \mathrm{e}^{-\mathrm{j}k\,|\,\mathbf{r}+(d/2)\,\mathbf{a}_x|} + I_2\, \mathrm{e}^{-\mathrm{j}k\,|\,\mathbf{r}-(d/2)\,\mathbf{a}_x|}\} \qquad (4.23)$$

where the subscript on d_H has been omitted for convenience.

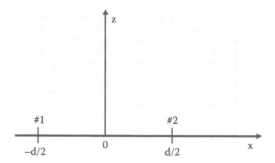

FIGURE 4.4 Two dipole antennas displaced in the horizontal plane.

Now, consider the quantities between the magnitude signs in Equation (4.23):

$$|\mathbf{r} + d_H/2\ \mathbf{a}_x| = [(r_x + d_H/2)^2 + r_y^2 + r_z^2\ ]^{1/2}$$

$$\simeq [r_x^2 + r_x\ d_H + r_y^2 + r_z^2]^{1/2}$$

$$= r[1 + (r_x\ d_H/r^2)]^{1/2}$$

$$\simeq r[1 + (d_H/2)\ (r_x/r^2)]$$

$$= r + (d_H/2)\ \sin\theta\ \cos\phi \qquad (4.24)$$

where we have assumed that $r_x \gg d_H$; that is, the distance to the far field observation point along the axis of displacement of the two antennas is much larger than the displacement itself. Also, we have used the identity $r_x = r\ \cos\phi\ \sin\theta$.

Proceeding in a manner similar to the derivation of Equation (4.24), one can show

$$|\mathbf{r} - d_H/2\ \mathbf{a}_x| = r - (d_H/2)\ \sin\theta\ \cos\phi \qquad (4.25)$$

Then, by substituting Equations (4.24) and (4.25) into Equation (4.23), one obtains the following expression for the far field produced by two z-oriented half-wave dipoles displaced from each other along the x-axis by a distance d_H:

$$E_\theta = [j\ Z_o\ f(\theta)\ e^{-jkr}/(2\pi\ r)]\ I_1\ e^{-jk(d/2)\ \sin\theta\ \cos\phi}$$

$$\{1 + (I_2/I_1)\ e^{jk\ d\sin\theta\ \cos\phi}\} \qquad (4.26)$$

where again the subscript on d_H has been omitted for convenience.

Let us now consider the special but still interesting case where antenna #2 is excited by a current having the same amplitude as the current exciting antenna #1 but with a different phase; that is,

$$I_2 = I_1 \, e^{j\xi} = I_i \, e^{j\xi} \tag{4.27}$$

Then, substituting Equation (4.27) into Equation (4.26), we get for the magnitude of the far field

$$|E_\theta| = [Z_o \, |I_i|/(2\pi \, r)] \, |f(\theta)| \, |\{1 + e^{j\zeta}\}| \tag{4.28}$$

where $\zeta = kd_H \sin\theta \cos\phi + \xi$.

Now, because $1 + e^{j\zeta} = 2 \, e^{j\zeta/2} \cos(\zeta/2)$, Equation (4.28) may be written in the following form:

$$|E_\theta| = [Z_o \, |I_i|/(2\pi \, r)] \, |f(\theta)| \, |A(\theta,\phi)| \tag{4.29}$$

where the array pattern function is

$$A(\theta,\phi) = 2 \cos(\zeta/2) = 2 \cos([kd_H \sin\theta \cos\phi + \xi]/2) \tag{4.30}$$

Note that in Equation (4.29), the magnitude of the far field is proportional to the product of the magnitude of the antenna element pattern function, $|f(\theta)|$, and the magnitude of the array pattern function, $|A(\theta,\phi)|$; this is a common result in antenna array analysis. Also, note in Equation (4.29) that as expected the radiation pattern is a function of ϕ and is not omnidirectional in the horizontal plane at $\theta = \pi/2$.

In the plane at $\theta = \pi/2$, $|f(\theta)| = 1$ and the far field expression simplifies to

$$|E_\theta (\theta = \pi/2)| = |E_\theta|_H = [\, Z_o \, |I_i|/(2\pi \, r)]$$

$$2 \, |\cos([kd_H \cos\phi + \xi]/2)| \tag{4.31}$$

The form of the pattern functions for the field in the horizontal plane represented by Equation (4.31) depend on two array parameters: the spacing between the antennas, d_H, and the phase difference between the antennas, ξ. The pattern functions will now be considered for two illustrative cases: the broadside array pattern and the endfire array pattern.

4.2.2 Broadside Arrays

To produce a radiation pattern that is maximum along the y-axis, transverse (or broadside) to the x-axis displacement between the antennas, and zero along the axis of displacement, one chooses the displacement to be half a wavelength and the phase difference between the antennas to be zero—that is, $d_H = \lambda/2$, $kd_H = (2\pi/\lambda)$ $(\lambda/2) = \pi$, and $\xi = 0$. Then Equation (4.31) becomes

$$|E_\theta|_H = [Z_o \, |I_i|/(2\pi \, r)] \, 2 \cos [(\pi/2) \cos\phi \,] \qquad (4.32)$$

The azimuthally varying part of this expression, $\cos [(\pi/2) \cos\phi]$, is tabulated in Table 4.1 for the values of azimuthal angle, $\phi = 0$, $\pi/2$, π, and $3\pi/2$. The corresponding radiation pattern is plotted in Figure 4.5.

4.2.3 Endfire Arrays

An entirely different radiation pattern can be obtained by changing the spacing or relative phase between the two antennas. An interesting alternative radiation pattern is one with a maximum along the positive x-axis (i.e., along the direction of the displacement between the antennas). Such a pattern is called an endfire pattern. It may be achieved by choosing the displacement to be one quarter of a wavelength and have antenna #2 lag antenna #1 by a phase difference of 90°. In other words, $d_H = \lambda/4$, $kd_H = (2\pi/\lambda)(\lambda/4) = \pi/2$, and $\xi = -\pi/2$. Then Equation (4.31) becomes

$$|E_\theta|_H = [Z_o \, |I_i|/(2\pi \, r)] \, 2 \cos [(\pi/4) \cos\phi - \pi/4] \qquad (4.33)$$

Then, for the present case of an endfire array, the azimuthally varying part of the expression for the far field magnitude, $\cos [(\pi/4) \cos\phi - \pi/4]$, is calculated with results displayed in Table 4.2

TABLE 4.1 Calculation of Azimuthal Variation of the Pattern Function of a Two-Element Broadside Array

ϕ	$(\pi/2) \cos\phi$	$\cos [(\pi/2) \cos\phi]$
0	$\pi/2$	0
$\pi/2$	0	1
π	$-\pi/2$	0
$3\pi/2$	0	1

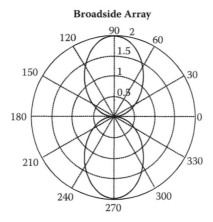

Broadside Array

FIGURE 4.5 Radiation pattern for a broadside array.

TABLE 4.2 Calculation of Azimuthal Variation of the Pattern Function of a Two-Element Endfire Array

ϕ	$(\pi/4)\cos\phi - \pi/4$	$\cos[(\phi/4)\cos\phi - \pi/4]$
0	0	1
$\pi/2$	$-\pi/4$	0.707
π	$-\pi/2$	0
$3\pi/2$	$-\pi/4$	0.707

for the values of azimuthal angle, $\phi = 0, \pi/2, \pi, 3\pi/2$. The corresponding radiation pattern is plotted in Figure 4.6.

The endfire maximum in the positive x-direction was achieved because antenna #1 is farther from the far field observation point and accumulates an extra phase retardation of $kd_H = \pi/2$ compared with the signal coming from antenna #2; however, because the signal from antenna #2 starts with a phase lag of $\pi/2$, the two signals arrive at the far field observation point in phase. Thus, they add (interfere constructively) and produce a maximum in the radiation pattern. Clearly, one can obtain an endfire radiation pattern with its maximum in the $-x$ direction by reversing the phase difference (i.e., by making $\xi = +\pi/2$).

4.2.4 Smart Antenna Arrays

In this section we have been considering the minimal case of an array of only two antennas. In general, one can deploy an array of many antennas. A flat-panel two-dimensional array of dipole antennas is sketched in Figure 4.7. As depicted, the antenna

Endfire Array

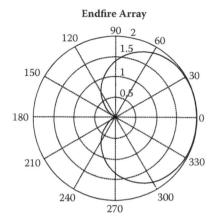

FIGURE 4.6 Radiation pattern for an endfire array.

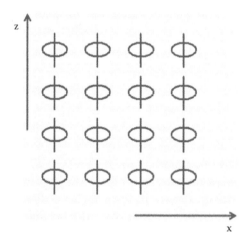

FIGURE 4.7 Flat-panel, two-dimensional array of dipole antennas.

wires are oriented along the z-axis, and there are many rows of antennas in the z-direction and many columns of antennas in the x-direction. One can expect such an array to combine the features of colinear arrays and arrays of antennas displaced in the horizontal plane; that is, the two-dimensional array would be expected to be able to control the radiation pattern in both the elevation and horizontal planes by appropriate phase control of the signals to the individual antenna elements. When such an array is used for receiving signals, it may be combined with signal processing techniques (viz., spatial filtering algorithms) to separate multiple signals on a single frequency channel based on their angle of arrival. The fabrication of two-dimensional arrays by microfabrication techniques (microstrip patch antennas) and the theory of operation of such arrays will be considered in Section 4.4.

4.3 Image Antennas

Before analyzing the operation of microstrip patch antennas, one needs to understand the concept of image antennas that is appropriately used to analyze radiation from an antenna in the presence of a conducting plane.

4.3.1 The Principle of Images

When one desires to derive the fields produced by an antenna in the presence of a conducting planar surface, one does not need to obtain a formal solution of the Helmholtz equation, which would include the currents on the antenna as well as all induced currents on the conducting surface as sources. Instead, one can obtain a simpler configuration to analyze, by replacing the conducting surface with an image antenna. The following procedure must be followed when making the substitution:

1. At the former location of the conducting plane, all fields must be the same as they were when the conductor was present.
2. The solution of the problem with the image antenna must be assumed to be valid only in the half-space containing the real antenna.

4.3.2 Quarter-Wave Monopole above a Conducting Plane

An important application of the method of images is to a monopole antenna, one quarter wavelength long, oriented with current flow in the vertical z-direction, and located above a conducting horizontal conducting plane. This type of antenna is driven at one end as shown in Figure 4.8. The current in the antenna wire is maximum at the drive point and goes to zero at the other end of the antenna. An electric field line produced by the monopole antenna is sketched in the figure. As required, the direction of the electric field at the conducting plane is normal to the plane.

Now, applying the principle of images, we remove the conducting plane and replace it by an image of the quarter-wave monopole on the side of the plane opposite from the real antenna as shown

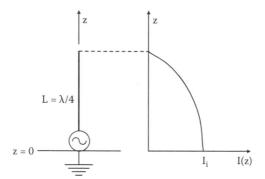

FIGURE 4.8   Quarter-wave dipole antenna above a conducting plane.

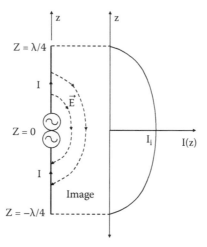

FIGURE 4.9   Conducting plate in Figure 4.8 replaced by an image antenna.

in Figure 4.9. By having the current in the image antenna flow in the same direction as in the real antenna, the resulting electric field will be normal to the plane that was formally occupied by the conductor as required by the first condition for applying the method of images. The real antenna and the image antenna together now form a half-wave dipole antenna. Thus, we have reduced the problem of a quarter-wave monopole above a conducting plane to the problem of a half-wave dipole in free space whose radiation fields are already known. Bear in mind however, that according to the second condition for applying the method of images, the field solutions for the half-wave dipole are only valid in the half-space above the plane that formerly was conducting and which contains the real antenna. There are no radiation fields in the half-space below the plane.

Thus, the total radiated power for the quarter-wave monopole above a conducting plane is one half the total radiated power for a half-wave dipole:

$$P_r = \frac{1}{2}\,(\frac{1}{2} \times 73.08\,|I_i|^2) = \frac{1}{2} \times 36.54\,|I_i|^2 = \frac{1}{2} \times R_r\,|I_i|^2 \qquad (4.34)$$

Equation (4.34) implies that the radiation resistance of a quarter-wave monopole above a conducting plane is $R_r = 36.54$ Ohms, compared with 73.08 Ohms for a half-wave dipole in free space.

An important related parameter is the directivity of the quarter-wave monopole above a conducting plane, which will be twice as large as the directivity of the half-wave dipole. This may be easily seen by recalling from Equation (3.49) that directivity

$$D = 4\pi r^2\,S_{max}/P_r$$

Because the fields in the upper half plane for the monopole will be the same as for the dipole, their values of S_{max} will also be the same. However, P_r for the monopole is one half of P_r for the dipole. Thus, it is clear from Equation (3.49) that the directivity of the monopole will be $D = 3.28$, twice as large as for the dipole.

4.3.3 Antennas for Handheld Cell Phones

It is apparent from the preceding section that quarter-wave monopole antennas above a conducting plane have advantageous directivity compared with a half-wave dipole if downward signal propagation is not required. Such is the case with handheld or vehicle-mounted cell phones that communicate with base station antennas that are typically mounted on high towers.

With a roof-mounted monopole antenna on a vehicle, the roof might serve the function of a conducting plane. However, no such plane is available with a handheld unit. Instead, a sleeve monopole antenna is used; such an antenna is sketched in Figure 4.10. Basically, the antenna consists of an exposed length of the center conductor of the coaxial feed cable and an electrically floating conducting sleeve. Currents, which produce the radiation fields, flow on both the coaxial inner conductor and on the sleeve. By adjusting the sleeve position relative to the tip of the inner conductor, radiation fields can be produced that are similar to those produced by a quarter-wave monopole antenna above a conducting plane.

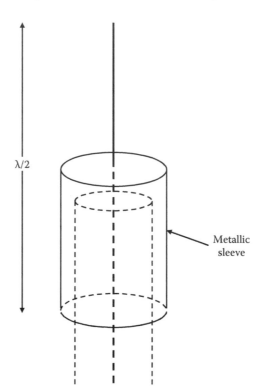

$\lambda/2$

Metallic
sleeve

FIGURE 4.10 Sleeve monopole antenna.

4.3.4 Half-Wave Dipoles and Reflectors

We will now consider a half-wave dipole antenna mounted near conducting planes, which effectively reduces the space occupied by the radiation fields and increases directivity by even a larger factor than that achieved with the monopole described in the preceding section.

First, consider a half-wave dipole mounted on the side of a metallic tower with its wire parallel to the tower surface. We will assume the tower diameter is large enough so that it can be modeled locally as a conducting plane. Using the principle of images, we replace the conducting plane by an image antenna located the same distance from the plane as the real antenna on a line normal to the plane, but on the opposite side of the plane as shown in Figure 4.11. The question that now arises is what will be the direction of current in the image antenna compared with the direction of current in the real antenna. This question is answered with reference to the requirement that "at the former location of the conducting plane, all fields must be the same as when the

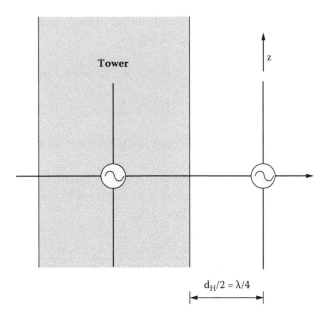

FIGURE 4.11 Half-wave dipole mounted on the side of a tower of large diameter.

conductor was present." This implies that at the former location of the conducting plane the parallel component of the electric field must vanish, which can only be satisfied if the direction of the current in the two antennas is opposite (i.e., there must be a relative phase difference between the current in the two antennas of 180°).

We now have two half-wave dipole antennas that are displaced in the horizontal plane by a distance d_H with a known relative phase shift between their exciting currents. Thus, to find the radiation fields in the half space containing the real antenna, we can use Equation (4.31). For example, if we choose d_H as half of a wavelength (i.e., $kd_H = \pi$), and with the phase shift $\xi = \pi$, we have for the far field in the horizontal plane:

$$|E_\theta|_H = [Z_o\,|I_i|/(2\pi\,r)]\,2\,|\cos\,([kd_H\cos\phi + \xi]/2)|$$

$$= [Z_o\,|I_i|\,/(2\pi\,r)]\,2|\cos\,([\pi\cos\phi + \pi]/2)| \qquad (4.35)$$

Along the positive x-axis ($\phi = 0$), the extra phase retardation of radiation from the image antenna is $kd_H = \pi$, while radiation from the real antenna has a phase lead of $\xi = \pi$. Thus, the signals from the two antennas will arrive at the far field observation point with a phase difference of 2π and will interfere constructively. As a result, this direction will be a direction of the maximum field that we calculate from Equation (4.34) as

$$|E_\theta|_{H,\,max} = [Z_o\,|I_i|/(2\pi\,r)]\,2\,|\cos\,([\pi + \pi]/2)|$$

$$= 2\,Z_o\,|I_i|/(2\pi\,r) \tag{4.36}$$

Let us now compare the directivity of two radiators:

1. Half-wave dipole antenna, which is positioned relative to a conducting plane as shown in Figure 4.11 with $d_H = \lambda/2$
2. Half-wave dipole in free space

The value of maximum field given by Equation (4.35) for radiator #1 is twice as large as the maximum field produced by radiator #2. Also, the maximum power density, S_{max}, which is proportional to the square of the field, will be four times larger for radiator #1 compared with radiator #2. From Equation (3.49), we note that directivity is directly proportional to S_{max} and inversely proportional to the total radiated power, P_r. Because only one half-wave dipole is fed for each of the radiators under consideration, we assume that for equal drive current, P_r stays about the same. Thus, the directivity of a half-wave dipole antenna, which is positioned relative to a conducting plane as shown in Figure 4.11 with stand-off distance $d_{H/2} = \lambda/4$, is four times larger than the directivity of a half-wave dipole in free space. In other words, the directivity of the dipole in the presence of the conducting plane is $D = 4 \times 1.64 = 6.56$.

Another interesting radiator is a half-wave dipole in the presence of corner reflectors. The arrangement is shown in the horizontal (x-y) plane in Figure 4.12. The real antenna is oriented so that current flows in the z direction normal to the page. Two conducting planes, one at $\phi = 45°$ and a second at $\phi = -45°$, meet to form a right angle; the surface of the planes are also normal to the page. To replace the conducting planes by image antennas, we first will need two image antennas, one for each of the planes.

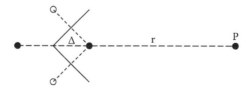

FIGURE 4.12 Half-wave dipole with a corner reflector.

Each of these image antennas will be located the same distance from the plane as the real antenna, on a line normal to the plane, but on the opposite side of the plane. Current in each of these image antennas will be in the opposite direction to the current in the real antenna as discussed in the previous example.

In the present case, one additional image antenna is required to make the parallel component of the electric field vanish in the corner. If the real antenna is located along the positive x-axis a distance Δ from the corner, the additional image will be located at $x = -\Delta$. Current in this additional image antenna will be in the same direction as current in the real antenna. It may now be seen from Figure 4.12 that there are four antennas equally spaced from the corner (one real antenna plus three image antennas). Two of the antennas carry current in the same direction as the real antenna, and two of the antennas carry current in the opposite direction. Thus, at the corner, the contributions to the parallel electric field from the four antennas will cancel as required.

Let us now consider the far field produced by each of the four antennas in the horizontal plane ($\theta = \pi/2$) and in the positive x direction ($\phi = 0$) under our usual assumption that the dimensions of the array are very small compared with the distance from the array to the observation point. The field due to the real antenna is

$$(E_\theta)_{H1} = j\, Z_o\, I_i\, e^{-jkr}/(2\pi\, r) \tag{4.37}$$

where r is the distance from the real antenna to the observation point.

The field due to the sum of the contributions from the two image antennas carrying current in a direction opposite from the current in the real antenna is

$$(E_\theta)_{H2} = 2\, j\, Z_o\, I_i\, e^{-j[k(r+\Delta)\, +\, \pi]}/(2\pi\, r) \tag{4.38}$$

The phase factor in Equation (4.38) includes the extra phase retardation due to displacement parallel to the x-axis by a distance Δ, and a phase shift of π due to the opposite sense of the current.

The remaining image antenna is displaced along the x-axis by a distance 2Δ compared with the real antenna, and its contribution to the far field is

$$(E_\theta)_{H3} = j\, Z_o\, I_i\, e^{-j[k(r+2\Delta)]}/(2\pi\, r) \tag{4.39}$$

Adding the contributions to the far field from Equations (4.37) to (4.39) gives for the total far field

$$(E_\theta)_H = [j\, Z_o\, I_i\, e^{-jkr}/(2\pi\, r)]\, [1 + 2\, e^{-j[k\Delta + \pi]} + e^{-j2k\Delta}] \quad (4.40)$$

The field given by Equation (4.40) can be maximized if we choose to make the distance from the real antenna to the corner as $\Delta = \lambda/2$. This makes $k\Delta = \pi$, and Equation (4.40) becomes

$$(E_\theta)_{H,\,max} = [j\, Z_o\, I_i\, e^{-jkr}/(2\pi\, r)]\, [1 + 2\, e^{-j2\pi} + e^{-j2\pi}]$$

$$= 4\, [j\, Z_o\, I_i\, e^{-jkr}/(2\pi\, r)]$$

Thus, we see that the maximum electric field produced by the half-wave dipole with a corner reflector can be four times larger than the maximum electric field produced by a half-wave dipole in free space. The corresponding maximum power density in the radiated wave can be 16 times larger, and the directivity $D = 16 \times 1.64 = 26.24$ (or 14.19 dBi). Thus, we see that the directivity of a dipole antenna in the presence of reflectors may be considerably larger than a dipole in free space. However, for values of directivity that are much larger still, interest has focused recently on arrays of microstrip patch antennas that will be described in the next section.

4.4 Rectangular Microstrip Patch Antennas

A very high gain, two-dimensional array of thousands of dipole antennas can be fabricated at relatively low cost using the techniques of optical lithography. Such "microstrip patch antenna arrays" are competitive with aperture antennas when high gain is required, and they are also electronically steerable as indicated in Section 4.2.4 on smart antenna arrays. Each element of the array consisting of a pair of dipoles is called a microstrip patch cavity. The construction and operation of a microstrip patch will be described below.

4.4.1 The TM$_{10}$ Microstrip Patch Cavity

The rectangular microstrip patch cavity shown in Figure 4.13 is constructed by fabricating a rectangular patch of conducting material (in the x-y plane with dimensions a and b) on top of a dielectric substrate layer, of thickness w, that in turn rests on

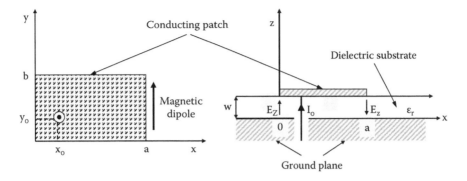

FIGURE 4.13 Layout of a rectangular microstrip patch antenna.

top of a conducting ground plane. The patch, together with the substrate and the ground plane, form a resonant cavity with open circuit boundary conditions at $x = 0$, $x = a$, $y = 0$, and $y = b$. The fields in the resonant cavity correspond to the TM_{10} mode at cutoff. This mode has an axial electric field component along the z-axis, E_z but no axial magnetic field component. The only nonzero transverse field component is H_y. Both E_z and H_y vary in the x-direction as shown in Figure 4.14, but do not vary along the y-axis or along the z-axis. At the open circuit boundaries at $x = 0$ and $x = a$, the tangential magnetic field vanishes while the magnitude of the tangential electric field is maximum. The direction of the electric field is opposite at the two sides of the cavity, so that $\mathbf{a}_n \times E$ has the same sign at the two open circuit "walls"; $\mathbf{a}_n$ is the unit vector normal to the outside surface of each "wall." The E_z fields along the edges at $x = 0$ and $x = a$ are the equivalent of two magnetic dipole antennas that are in phase. As shown in Figure 4.13, the center conductor of a coaxial cable may be inserted through the ground plane to excite the cavity.

The fringing and radiation electric fields are depicted in Figure 4.15. As suggested by R.E. Collin (*Antennas and Radiowave Propagation*, McGraw-Hill, New York, 1985), each edge looks roughly like a slot in a metallic wall with slot dimensions b in the y-direction and w in the z-direction. The power radiated by such a patch edge "slot" is given by

$$P_{pe} = \tfrac{1}{2}(|E_0|^2\, w^2/R_{pe}) \qquad (4.41)$$

where R_{pe} is the radiation resistance of the patch edge, and E_o is the maximum magnitude of the cavity electric field.

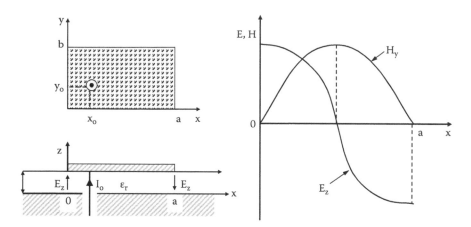

FIGURE 4.14 E_z and H_y fields in a microstrip patch cavity.

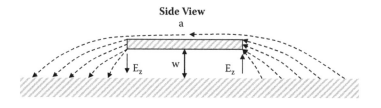

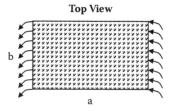

FIGURE 4.15 Fringing and radiation electric fields.

4.4.2 Duality in Maxwell's Equations and Radiation from a Slot

To find the value of the patch edge radiation resistance, Collin invokes the similarity between a slot with an electric field across the short dimension and a conducting strip with a transverse magnetic field across the short dimension. These two entities with the same dimensions (viz., w for the short dimension and b for the long dimension) are compared in Figure 4.16. The conducting strip on the right has a current I_o flowing along its length and a corresponding magnetic field H_o across its short dimension. The slot on the left side of the figure has an electric field E_o across its

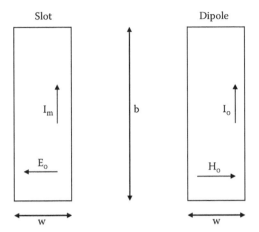

FIGURE 4.16 Comparison of a slot in a conducting plate and dipole in free space.

short dimension and may be imagined to have a fictional magnetic current I_m flowing along its length.

Now it should be recognized that if $b \gg w$, the conducting strip is a dipole antenna for which radiated power is

$$P_d = \tfrac{1}{2} \, |I_o|^2 \, R_d \tag{4.42}$$

where R_d is the radiation resistance of the dipole antenna.

Now, to relate I_o to H_o, we note that the surface current, J_s, is equal to H_o and that it flows on both sides of the conducting strip (i.e., through a width of $2w$). Thus,

$$I_o = 2w \, J_s = 2w \, H_o \tag{4.43}$$

Combining Equations (4.42) and (4.43), the power radiated by the conducting strip (i.e., the dipole antenna) can be related to the magnetic field at its surface as

$$P_d = \tfrac{1}{2} |2w \, H_o|^2 \, R_d = 2w^2 \, |H_o|^2 \, R_d \tag{4.44}$$

Maxwell's equations include a high degree of symmetry between the electric and magnetic fields (often referred to as duality). This duality implies that the power radiated by the slot on the left side of Figure 4.16, P_{pe}, will be equal to the power radiated by the conducting strip shown on the right side of the figure, P_d, if

$$|E_o| = |H_o| \, Z_o \qquad (4.45)$$

Thus, from Equations (4.44) and (4.45),

$$P_{pe} = 2w^2 \, |E_o|^2 \, R_d \, /Z_o^2 \qquad (4.46)$$

Now, comparing Equation (4.46) with Equation (4.41), we see that the patch edge radiation resistance, R_{pe}, may be expressed in terms of the radiation resistance of a dipole antenna with the same dimensions as

$$R_{pe} = Z_o^2/4 \, R_d \qquad (4.47)$$

4.4.3 Radiation from the Edges of a Microstrip Cavity

The radiation fields produced by a slot and by a patch edge are compared in Figure 4.17. It may be seen that the radiation from the patch edge is similar to the radiation from the slot, but it only fills the half space in front of the patch antenna and its ground plane, and there is an image antenna in the ground plane. The situation is similar to the case of a dipole antenna alongside a conducting plane that was examined in Section 4.3.4, but now, with the distance between the antenna and the conducting plane almost zero. The total radiated power is not changed by the presence of the conducting plane, but the directivity is increased by a factor of four compared with the slot. For example, if the slot is much shorter than the wavelength, its directivity will be the same as a Hertzian dipole—that is, $D_{slot} = 1.5$; the directivity of a patch edge with the same length will be $D_{pe} = 4 \times 1.5 = 6$.

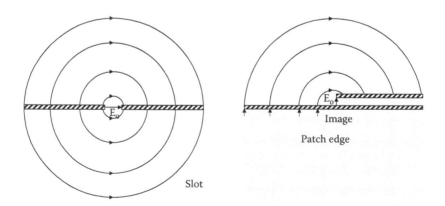

FIGURE 4.17 Comparison of radiation fields from a slot and from a patch edge.

We will now consider the resonant cavity formed by the microstrip patch, the dielectric substrate of relative permittivity ε_r, and the ground plane as shown in Figure 4.18. The cavity operates in the TM_{10} mode at cutoff; that is,

$$k^2 = \omega^2 \, \varepsilon_r \varepsilon_o \mu_o = (\pi/a)^2$$

or

$$a = \lambda_o/(2 \, \varepsilon_r^{\frac{1}{2}}) \tag{4.48}$$

With choice of the center frequency at which the microstrip patch antenna is to operate and with a dielectric substrate of known permittivity, the width of the patch is determined by Equation (4.48).

The fields in the cavity are

$$\boldsymbol{E} = E_o \cos (\pi \, x/a) \, \boldsymbol{a}_z \tag{4.49a}$$

and

$$\boldsymbol{H} = (j \, \varepsilon_r^{\frac{1}{2}} \, E_o/Z_o) \sin (\pi \, x/a) \, \boldsymbol{a}_y \tag{4.49b}$$

The energy stored in the cavity may be calculated using the fields in Equation (4.49):

$$W = \frac{1}{2} \iiint \varepsilon \left| E \right|^2 dV = \frac{1}{2} \left| E_o \right|^2 \varepsilon_r \varepsilon_o bw \int_0^a \cos^2 (\pi x/a) dx$$

$$= \frac{1}{4} \left| E_o \right|^2 \varepsilon_r \varepsilon_o \, b \, w \, a \tag{4.50}$$

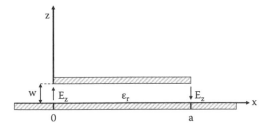

FIGURE 4.18 Resonant cavity formed by a microstrip patch.

The power radiated from the two edges of the patch at $x = 0$ and $x = a$ may be obtained by using Equation (4.47) as

$$P_r = 2\,P_{pe} = 4w^2\,|E_o|^2\,R_d/Z_o^2 \tag{4.51}$$

Next, we can consider the cavity loaded quality factor, Q. One reason quality factor is important is that it determines width of the resonant frequency response of the cavity, or equivalently, the effective operating bandwidth of the antenna, according to

$$\Delta f/f_o = 1/Q \tag{4.52}$$

where Δf is the 3 dB width of the resonance centered at the frequency f_o.

Q depends on the stored energy in the cavity, W, and on the power lost from the cavity, P_{loss}, as given by

$$Q = 2\pi\,f_o\,W/P_{loss} \tag{4.53}$$

Power loss may be produced by a number of physical mechanisms as follows:

1. Radiation from the patch edges producing a radiated power loss, P_r, as given by Equation (4.50)
2. Power absorption by the dielectric substrate, P_d
3. Power absorption in the conducting patch material and in the conducting ground plane, P_c
4. Power lost by coupling into the feed line, $P_{coupling}$

The first three factors are combined to give the unloaded quality factor, Q_o, which is defined by

$$Q_o = 2\pi\,f_o\,W/(P_r + P_d + P_c) \tag{4.54}$$

The strength of the coupling is usually adjusted so that

$$P_{coupling} = P_r + P_d + P_c \tag{4.55}$$

The condition in Equation (4.55) is called critical coupling. It ensures that a signal traveling along the feedline will see the cavity as a matched load and not be reflected. Clearly, with critical coupling,

$$Q = \tfrac{1}{2} Q_0 \qquad (4.56)$$

Expressions for Q_d and Q_c can be found in the book by Collin (*Antennas and Radiowave Propagationi*, McGraw-Hill, New York, 1985). The expression for Q_d is

$$Q_d = \varepsilon'/\varepsilon'' \qquad (4.57)$$

where ε' and ε'' are, respectively, the real and imaginary parts of the permittivity of the dielectric substrate (i.e., $\varepsilon''/\varepsilon'$ is the loss tangent).

The expression for the quality factor due to conductor loss in a microstrip patch cavity operating in the TM_{10} mode is

$$Q_c = (1/\pi) f_o \varepsilon_r \varepsilon_o \sigma \delta_s w (k_o a Z_o)^2 \qquad (4.58)$$

where σ is the conductivity of the patch and ground plane material, and δ_s is the skin depth in that material at the operating frequency.

Example of Microstrip Patch Design Calculations

As an example of a microstrip patch antenna, consider one operating at a center frequency of 3 GHz ($\lambda_o = 10$ cm) with patch height $b = \lambda_o/5 = 2$ cm. The dielectric substrate has thickness $w = 1$ mm, relative permittivity $\varepsilon_r = 4$, and a dielectric breakdown field of $E_{max} = 6 \times 10^6$ V/m. Calculate the following antenna parameters:

1. The width, a, of the patch antenna
2. The maximum radiated power up to the dielectric breakdown limit
3. The bandwidth of the patch antenna assuming that the dominant power loss from the cavity is by radiation and that the cavity is critically coupled to its feedline

The width of the patch is easily found from Equation (4.47) as

$$a = \lambda_o/(2 \varepsilon_r^{1/2})$$

$$= 10 \text{ cm}/(2 \times 4^{1/2})$$

$$= 2.5 \text{ cm}$$

The maximum radiated power is found from Equation (4.50) by setting $E_o = E_{max}$ to obtain

$$P_r(max) = 4w^2 |E_{max}|^2 R_d /Z_o^2 \qquad (4.59)$$

In order to evaluate this equation we must know R_d, the radiation resistance of a dipole antenna with the same dimensions as the patch edge. Because the length of the patch edge is small compared with a wavelength and because the magnetic current along its length is constant, it is appropriate to use the expression for the radiation resistance of a Hertzian dipole as given by Equation (3.42):

$$R_d = 80 \ \pi^2 \ b^2/\lambda_o^2$$

$$= 80 \ \pi^2 \ 2^2/10^2$$

$$= 32 \ Ohms$$

Then, using this value in Equation (4.59) yields the maximum power that can be radiated from the patch:

$$P_r(max) = 4 (0.001)^2 |6 \times 10^6|^2 \ 32/(120 \ \pi)^2 \ Watts$$

$$= 3.2 \times 10^4 \ Watts$$

To evaluate the bandwidth, Δf, we begin by noting that the dominant loss of power from the cavity is due to radiation (i.e., $Q_o \approx 2\pi f_o \ W/P_r$).

Using Equations (4.49) and (4.50), we can estimate the unloaded quality factor of the cavity as

$$Q_o \approx 2\pi \ f_o \ W/P_r. = 2\pi \ f_o \ \varepsilon_r \varepsilon_o \ b \ a \ Z_o^2/(16 \ w \ R_d)$$

$$= (2\pi \times 3 \times 10^9 \times 4 \times 8.85 \times 10^{-12} \times 0.02 \times 0.025 \times 377^2)/(16 \times 0.001 \times 32)$$

$$= 29$$

Then, from Equations (4.51) and (4.56), the bandwidth is approximately

$$\Delta f \approx 2 \ f_o/Q_o = 2 \times 3 \times 10^9/29 \approx 2 \times 10^8 \ Hz$$

4.4.4 Array of Microstrip Patch Antennas

Often, microstrip patch antennas are fabricated as a large array of patches on a planar sheet, and very large values of gain are obtainable. The total gain is given by

$$G = \eta \, D_1 \, D_A \qquad (4.60)$$

where η is antenna array efficiency and depends on such factors as dielectric and conductor losses. D_1 is the directivity of one patch edge (i.e., one magnetic dipole antenna). D_A is the array directivity factor.

From the preceding discussion in this section, we can estimate that $D_1 \approx 6$. Also, from the discussion of the directivity of N coherently excited antennas in Section 4.1.3, we set $D_A = N$. Thus, Equation (4.60) for an array of N patch edge antennas with coherent excitation becomes

$$G \approx 6 \, N \eta \qquad (4.61)$$

Example of Calculations for an Array of Microstrip Patches

Let us now make a calculation for an array of 32 × 32 patches. Because there are two antennas per patch, $N = 2$ × 32 × 32 = 2048, and directivity $D = D_1 \, D_A \approx 6 \times 2048$ = 12,300 or 41 dBi. The gain of a 32 × 32 array operating at a frequency of 34.8 GHz was calculated by Bhartia and Bahl (*Millimeter Wave Engineering and Applications*, John Wiley & Sons, New York, 1984) taking losses into account; the calculated value was $G = 38$ dBi, which is in reasonable agreement with our estimate that directivity $D \approx 41$ dBi.

The experimental measurement of gain for the 32 × 32 array of patches operating at 38.4 GHz was 29 dBi. The smaller experimental number was attributed largely to nonuniformity in the thickness of the dielectric substrate that was especially important at the high operating frequency, where the height of the nonuniformity might not be small compared with the wavelength.

In summary, microstrip patch antenna arrays are capable of achieving very high gain. They are easily fabricated at relatively low cost, have a low profile, and are flexible. They could be wrapped around a cylinder (a ship smokestack or an airplane fuselage, for example). Their fractional bandwidth, however, is only modest.

Problems

4.1. A colinear array of half-wave dipoles is to be used to produce an omnidirectional radiation pattern with a directivity of at least 14 dBi at a frequency of 900 MHz. Calculate the minimum overall length of the array and the 3 dB beamwidth, given that the center-to-center spacing between adjacent antennas is 0.5 wavelengths.

4.2. If the 14 dBi antenna array described in Problem 4.3 is mounted on top of a 50 m tall tower, calculate the phase advance across the entire length of the array to ensure that the outer edge of the 3 dB beamwidth hits the ground at a distance of 1 km from the tower.

4.3. For the array and phase advance in Problem 4.2, calculate the distance from the tower in which nulls may occur in the radiation pattern.

4.4. Two vertically polarized dipole sources of equal amplitude but opposite phase are separated by a distance of $1/2$ in the horizontal plane. Sketch the array radiation pattern in the horizontal plane containing the sources.

4.5. Consider a 90° corner reflector antenna with a half-wave dipole as the driving element and an element to corner spacing of $\lambda/2$. Calculate the directivity.

4.6. Repeat the calculation in Problem 4.5 for an element to corner spacing of λ.

4.7. A microstrip patch antenna is made from a rectangular patch with $a = 2$ cm, $b = 1/2$, and $w = 0.2$ cm. The substrate has a relative permittivity of 2.5. Find the resonant frequency of the cavity for the TM_{10} mode.

4.8. For the patch antenna in Problem 4.7, assume that copper and dielectric losses are negligible. Find the Q and the 3 dB bandwidth of the cavity. Hint: For a dipole antenna of length b, with *uniform* current along its length, the radiation resistance R_d is given by $R_d^{-1} = \{[\lambda/b]^2 + [4/3] [\lambda/(b + \lambda)]\}/(80\pi^2)$.

4.9. For the patch antenna described above, assume that the maximum electric field in the substrate is 9 kV/cm. Calculate the maximum radiated power.

4.10. If the microstrip patches described above are arrayed as a 100 × 100 array of patches, calculate the gain of the array and the maximum power density at a distance of 10 km.

4.11. What is the effective area of a parabolic reflector antenna that would have the same directivity as the microstrip patch array in Problem 4.10? Discuss how the reflector area compares to the area occupied by the microstrip patch array. [Assume that the horizontal spacing between patches is chosen for equal horizontal spacing between all the radiating patch edges, and that vertical spacing is chosen to make the distance between the top of one patch and the bottom of the neighboring patch equal to one wavelength.

Bibliography

1. P. Bhartia and I. J. Bahl, *Millimeter Wave Engineering and Applications* (John Wiley & Sons, New York, 1984), 598–611.
2. R.E. Collin, *Antennas and Radiowave Propagation* (McGraw-Hill, New York, 1985), 104–120, 261–286.
3. W.L. Sturztzman and G. A. Thiele, *Antenna Theory and Design* (John Wiley & Sons, New York, 1981), 108–144.

CHAPTER 5

Radio Frequency (RF) Wave Propagation

The Friis transmission formula, which was derived in Chapter 3, implies a simple model of propagation loss simply due to a spherical wave spreading out from its source without obstruction. Although this model may have applicability to satellite communications (SATCOM), where the satellite transmitter is radiating into free space, it cannot apply to propagation paths full of obstructions, such as buildings and the earth. Then, a number of physical phenomena will come into play that will influence the propagation loss, including reflection, refraction, absorption, multipath phase interference, and diffraction. The Friis formula for free space propagation predicts a path loss that is proportional to the square of the frequency times the square of the distance between transmitter and receiver (i.e., the range). In contrast to the Friis formula, path loss empirically measured for cell phone communications in urban areas depends on the fourth power of the range and shows the parametric dependence given by the following expression:

$$L_{empirical} \sim \frac{r^4 f^2}{h_b^2 h_m}$$ (5.1)

where r is the distance between transmitter and receiver (i.e., the range), f is the frequency, h_b is the height of the base station antenna, and h_m is the height of the mobile antenna.

In this chapter, we will begin by considering the simple case of free space propagation and then add other physical mechanisms that affect the path loss until a relationship is arrived at that is similar to Equation (5.1).

5.1 Some Simple Models of Path Loss in Radio Frequency (RF) Wave Propagation

5.1.1 Free Space Propagation

For antennas in free space, the ratio of signal power received to transmitter power is given by the Friis formula, Equation (3.79):

$$\frac{P_s}{P_t} = \frac{G_t G_r}{(4\pi r / \lambda)^2} \tag{5.2}$$

This equation can be compared with the general communication system equation, Equation (1.16), which has been specialized for the case of zero loss in the feeder cables:

$$P_s = \frac{P_t \, G_t \, G_r}{L} \tag{5.3}$$

It may be seen that the free space path loss is

$$L = L_F = \left(\frac{4\pi r}{\lambda}\right)^2 \tag{5.4}$$

or

$$L_F = \left(\frac{4\pi r f}{c}\right)^2 \tag{5.5}$$

Equivalently, free space path loss may be expressed in decibels as

$$L_F(\text{dB}) = 32.4 + 20 \log r_{km} + 20 \log f_{MHz} \tag{5.6}$$

In Equation (5.6), as indicated by the subscripts, the range is normalized to 1 km, and the frequency is normalized to 1 MHz.

Comparing L_F with the empirical path loss in Equation (5.1), we see that the dependence on range is much weaker in the free space model, and there is no dependence on antenna heights above the earth, because the earth is not present in the free space model. We will consider the effect of the presence of the earth in the remainder of Section 5.1, beginning with a review of reflection and refraction phenomena that might occur when an RF wave propagation path is incident on the earth.

5.1.2 Laws of Reflection and Refraction at a Planar Boundary

Snell's laws give the direction of the reflected and transmitted waves. When a plane wave is incident on a planar boundary between two dielectrics, it will be partly reflected and partly transmitted. If the angle between the wavevector of the incident wave and the normal to the boundary is the angle of incidence, θ_i, as shown in Figure 5.1, then the angle between the reflected wavevector and the normal to the boundary is given by Snell's Law of Reflection as simply

$$\theta_r = \theta_i \tag{5.7}$$

Equation (5.7) holds even when the two media are conductive.

The angle between the wavevector of the transmitted wave and the normal to the boundary is called the angle of transmission, θ_t.

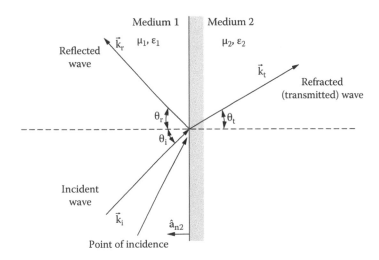

FIGURE 5.1   Geometry for wave reflection and transmission at an interface.

When the media have zero conductivities, θ_t obeys Snell's Law of Refraction:

$$\frac{\sin \theta_i}{\sin \theta_t} = \frac{\sqrt{\varepsilon_2 \mu_2}}{\sqrt{\varepsilon_1 \mu_1}} \qquad (5.8)$$

where the subscripts 1 and 2 refer to the medium of incidence and the medium of transmission, respectively.

Equation (5.8) will hold approximately if the conductivities of the two media are small (i.e., $\sigma \ll \omega\varepsilon$).

The relative magnitudes of the fields in the reflected and transmitted waves are given by the Fresnel coefficients that are defined in Equations (5.9a) through (5.9d). The coefficients depend on whether the incoming wave is polarized parallel to or normal to the plane of incidence. The plane of incidence is defined as containing the wavevector of the incident wave and the normal to the interface.

For an incoming wave with an electric field parallel to the plane of incidence, the magnitude of the electric field in the reflected wave normalized to the magnitude of the incident electric field is

$$R_{\parallel} = \frac{E_{r\parallel}}{E_{i\parallel}} = \frac{Z_2 \cos \theta_t - Z_1 \cos \theta_i}{Z_2 \cos \theta_t + Z_1 \cos \theta_i} \qquad (5.9a)$$

and the normalized magnitude of the electric field in the transmitted wave is given by

$$T_{\parallel} = \frac{E_{t\parallel}}{E_{i\parallel}} = \frac{2Z_2 \cos \theta_i}{Z_2 \cos \theta_t + Z_1 \cos \theta_i} \qquad (5.9b).$$

For an incoming wave with an electric field normal to the plane of incidence, the magnitude of the electric field in the reflected wave normalized to the magnitude of the incident electric field is

$$R_{\perp} = \frac{E_{r\perp}}{E_{i\perp}} = \frac{Z_2 \cos \theta_i - Z_1 \cos \theta_t}{Z_2 \cos \theta_i + Z_1 \cos \theta_t} \qquad (5.9c)$$

and the magnitude of the electric field in the transmitted wave is

$$T_\perp = \frac{E_{t\perp}}{E_{i\perp}} = \frac{2Z_2 \cos\theta_i}{Z_2 \cos\theta_i + Z_1 \cos\theta_t} \tag{5.9d}$$

In Equation (5.9), the wave impedance in each medium is defined in general as a complex quantity:

$$Z = [\, j\,\omega\mu/(\sigma_c + j\omega\varepsilon)]^{1/2} \tag{5.10}$$

It should be noted from Equation (5.9a) that for grazing incidence (i.e., where the angle of incidence approaches 90°), the reflection coefficient for parallel polarization, $R_\parallel \Rightarrow 1$.

Similarly, from Equation (5.9c), at grazing incidence, the reflection coefficient for perpendicular polarization $R_\perp \Rightarrow -1$.

In Figure 5.2, a plot of the Fresnel coefficients taken is shown for the case of a wave incident from dry air onto very dry earth ($\varepsilon_r = 2.53$, $\sigma \approx 0$) as a function of the angle of incidence. Note that at $\theta_i = 57.8°$ the reflection coefficient for parallel polarization vanishes; this angle is called the polarizing angle or Brewster angle, θ_B.

The Brewster angle satisfies the relationship

$$\theta_B = \tan^{-1}(\varepsilon_2/\varepsilon_1)^{1/2} \tag{5.11}$$

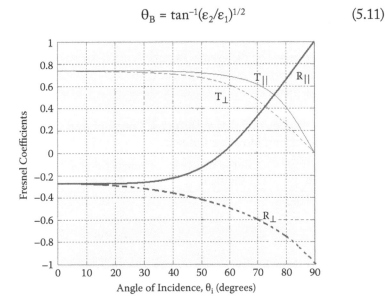

FIGURE 5.2 Plot of Fresnel coefficients vs. θ_i. [When medium #1 is dry air ($\varepsilon_1 = \varepsilon_0$, $\mu_1 = \mu_0$, $\sigma_1 = 0$), and medium #2 is dry earth ($\varepsilon_2 = 2.53\varepsilon_0$, $\mu_2 = \mu_0$, $\sigma_2 = 0$).]

When the angle of incidence equals the Brewster angle, only the component of the wave with perpendicular polarization is reflected (i.e., with an incoming randomly polarized wave, a linearly polarized reflected wave would result).

Also, note from Figure 5.2, that at grazing incidence, $R_\| \Rightarrow 1$ and $R_\perp \Rightarrow -1$ as expected from our previous discussion.

The dielectric constant and the conductivity of the earth depend on its moisture content as indicated in Table 5.1.

It may be seen from Table 5.1 that the values of ε_r and of σ both increase as the water content of the earth increases. This is hardly surprising because for seawater, $\varepsilon_r = 81$ and $\sigma = 5$ S/m. One important matter to remember is that for all types of earth, as grazing incidence is approached, the reflection coefficient for parallel polarization, $R_\| \Rightarrow 1$, and the reflection coefficient for perpendicular polarization $R_\perp \Rightarrow -1$.

5.1.3 Effect of Surface Roughness

The preceding section assumes that the interface between the two media is planar and smooth. In practice, for example, when a signal is reflected by the ground, the interface will consist of small hills and valleys. In this section, we consider what degree of undulation in the interface may be tolerated without making the relationships in the preceding section inapplicable.

Let us consider reflection from two regions on the undulating surface that differ in height by an amount Δ_h as shown in Figure 5.3. The difference in path length for waves reflected from the two regions is $2\Delta_h \cos\theta_i$, and the corresponding phase difference is given by

TABLE 5.1 Values of Relative Dielectric Constant and Conductivity of Earth with Varying Moisture Content

Type of Earth	Relative Dielectric Constant, ε_r	Conductivity, σ (S/m)	Source
Very dry, Saharan	2.53	~0	Sihvola, 1999
"Dry"	4–7	0.001	ITU, 1992
"Average"	15	0.005	ITU, 1992
"Wet"	25–30	0.02	ITU, 1992

Sources: International Telecommunications Union (ITU), ITU-R Recommendation 527-3, "Electrical Characteristics of the Surface of the Earth," Geneva, 1992; A.H. Sihvola, *Electromagnetic Mixing Formulae and Applications* (Institution of Engineering and Technology, IET, London, 1999). See page 278 for permittivity of very dry earth. With permission.

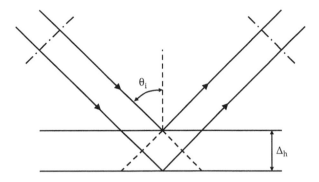

FIGURE 5.3 Wave reflection from a rough surface.

$$\Delta_\alpha = 2\,k\,\Delta_h\,\cos\theta_i \qquad\qquad (5.12)$$

A generally accepted criterion for considering a surface to be smooth would be that its characteristic height deviation produces a phase variation in the reflected wave smaller than $\pi/8$ radians. Using Equation (5.12), this criterion would imply that in order for a surface to be considered smooth, the standard deviation of its height, σ_h, must satisfy the inequality

$$\sigma_h < \lambda/(32\,\cos\theta_i) \qquad\qquad (5.13)$$

Note that the inequality (5.13) becomes easier to satisfy as grazing incidence is approached (i.e., as θ_i approaches 90°).

5.1.4 Plane Earth Propagation Model

We are now in a position to consider a propagation model that includes the effect of reflection from the earth and includes as parameters the heights of the antennas above the earth. There are two antennas, the base station antenna with height h_b, and the mobile antenna with height h_m. We will assume that the heights of the antennas are small compared with the distance between the antennas (i.e., $h_b \ll r$ and $h_m \ll r$), so that near grazing incidence of radiation on the earth is expected. In line with the assumption of near grazing incidence, we will assume that inequality (5.13) is satisfied, and the reflecting surface appears smooth. The communication link is drawn in Figure 5.4.

As shown in Figure 5.4, there are two paths for a signal from the base station antenna to reach the mobile antenna (i.e., the direct path designated r_1 and the path involving reflection from the earth designated r_2). The lengths of these paths are given by

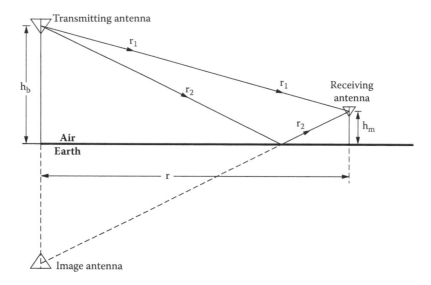

FIGURE 5.4 Geometry for a plane earth propagation model.

$$r_1 = \sqrt{(h_b - h_m)^2 + r^2} \approx r\left[1 + \tfrac{1}{2}(h_b - h_m)^2/r^2\right] \quad (5.14a)$$

and

$$r_2 = \sqrt{(h_b + h_m)^2 + r^2} \approx r\left[1 + \tfrac{1}{2}(h_b + h_m)^2/r^2\right] \quad (5.14b)$$

Then the difference in path length is easily calculated to be

$$r_1 - r_2 = \frac{2h_m h_b}{r} \quad (5.15)$$

The amplitude of the electric field at the mobile antenna is then

$$E_{total} = E_{direct} + E_{refl.} = E_{direct}\left|1 + R\exp(-j\frac{k2h_b h_m}{r})\right| \quad (5.16)$$

where R is the reflection coefficient for the wave on path #2 as it encounters the earth.

Equation (5.16) can be readily evaluated using the near-grazing values of reflection coefficients (i.e., $R_\perp \approx -1$ and $R_\parallel \approx 1$).

Equation (5.16) can reduce to an especially simple form for the case of perpendicular polarization and in the limit $2kh_b h_m/r \ll 1$. Then the exponential in Equation (5.16) can be approximated as

$$\exp\left(-j\frac{k2h_b h_m}{r}\right) \approx 1 - j\frac{k2h_b h_m}{r}$$

and Equation (5.16) reduces to

$$E_{total} = E_{direct}\frac{2kh_b h_m}{r} \qquad (5.17)$$

Because power density is proportional to the square of the electric field magnitude, Equation (5.17) implies that the received signal power is

$$P_{s,total} = P_{s,direct}\frac{(2k)^2 h_b^2 h_m^2}{r^2} = \frac{P_t G_t G_r}{L_t L_r}\frac{(2k)^2 h_b^2 h_m^2}{L_F r^2} \qquad (5.18)$$

where we have used $P_{s,direct} = \dfrac{P_t G_t G_r}{L_t L_r L_F}$.

Thus, in the presence of reflection from the plane earth, the new value of path loss, $L_{p.e.}$, becomes

$$L_{p.e.} = L_F\frac{r^2}{(2k)^2 h_b^2 h_m^2} = \left(\frac{4\pi r f}{c}\right)^2\frac{r^2 c^2}{(4\pi f)^2(2k)^2 h_b^2 h_m^2}$$

or

$$L_{p.e.} = \frac{r^4}{h_b^2 h_m^2} \qquad (5.19)$$

Equation (5.19) may be compared with the empirical expression for path loss in a cell phone system in an urban area given in Equation (5.1). We see that both expressions have dependence on range as r^4, and both depend inversely on the heights of the antennas. However, the plane earth path loss is independent of frequency while the empirical path loss varies as f^2. Thus, not surprisingly, other physical processes must be accounted for to achieve a propagation model that matches the empirical result for RF wave propagation in an urban area; chief among these physical processes is diffraction.

5.2 Diffraction over Single and Multiple Obstructions

5.2.1 Diffraction by a Single Knife-Edge

Diffraction is the phenomenon that allows electromagnetic waves to bend around objects that block the direct path between the transmitting antenna and the receiving antenna. It is hardly surprising that diffraction plays a large role in cell phone communications in urban areas. The mobile unit may be at the bottom of a canyon-like street with tall buildings on either side, while the base station antenna may be on a tower many streets away; it is rather difficult to imagine a direct unblocked path.

The basic concept describing the phenomenon of diffraction is contained in principles proposed by Christian Huygens (1629 to 1695) and Augustin Fresnel (1788 to 1827). Huygens' principle states that

> The wavefront of a propagating wave of light at any instant conforms to the envelope of spherical wavelets emanating from every point on the wavefront at the prior instant.

Fresnel's elaboration of Huygens' principle states that

> The amplitude of the wave at any given point equals the superposition of the amplitudes of all the secondary wavelets at that point.

Together, these two statements are referred to as the Huygens–Fresnel principle. Although the original statements referred to light waves, Gustav Kirchhoff (1824 to 1887) subsequently showed that they could be deduced from Maxwell's equations, and so pertained to all electromagnetic waves.

In Figure 5.5, an absorbing screen with a flat, sharp top (knife-edge) is shown to be blocking the direct ray path between a transmitting antenna and a receiving antenna. The wavefronts before and after the screen are also sketched according to the

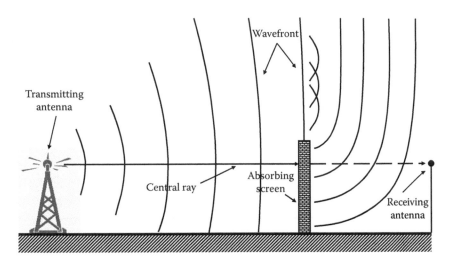

FIGURE 5.5   Propagation into the shadow region behind an absorbing screen.

Huygens–Fresnel principle. It is seen that some wave power still reaches the receiving antenna, even though it is deep within the shadow region of the absorbing screen.

The fraction of wave power reaching the receiver has been analyzed using a mathematical formulation based on the Huygens–Fresnel principle. The result states that the received power is the same as would be calculated in the absence of the screen except that the path loss, L(dB), is greater than the free space path loss, L_F(dB), by the addition of an excess path loss due to the presence of the absorbing screen with its knife-edge, L_{ke}(dB); that is,

$$L(dB) = L_F(dB) + L_{ke}(dB) \tag{5.20}$$

The basic geometry, on which the calculation of L_{ke}(dB) is based, is shown in Figure 5.6.

The important dimensions are

  d'_1, the distance from the transmitter to the screen along the
    ray path
  d'_2, the distance from the screen to the receiver along the direct
    ray path
  h'_e, the excess height of the top of the screen with respect to the
    direct path (h'_e can have a negative value when the top of the
    screen is below the direct path)

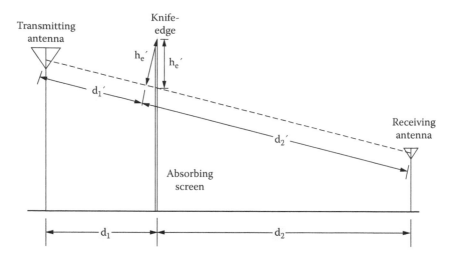

FIGURE 5.6 Geometry for diffraction by a single knife-edge.

When h'_e is much smaller in magnitude than either d'_1 or d'_2, to good approximation, one can make the calculations by using the vertical projection of h'_e, which is denoted by the unprimed symbol h_e, and by the horizontal projections of d'_1 and d'_2, which are denoted by d_1 and d_2, respectively.

The excess path loss depends on the particular combination of these dimensions known as the diffraction parameter, v, which is defined by

$$v = h'_e \sqrt{\frac{2(d'_1 + d'_2)}{\lambda d'_1 d'_2}} \approx h_e \sqrt{\frac{2(d_1 + d_2)}{\lambda d_1 d_2}} \qquad (5.21)$$

The excess path loss is expressed in terms of the diffraction parameter by the following expression:

$$L_{ke}(v) = -20 \log|F(v)| \qquad (5.22)$$

where

$$|F(v)| = \tfrac{1}{2}[\tfrac{1}{2} + C^2(v) - C(v) + S^2(v) - S(v)] \qquad (5.23)$$

$C(v)$ and $S(v)$ are, respectively, the Fresnel cosine integral and the Fresnel sine integral that are tabulated in optics textbooks such as that by Jenkins and White (*Fundamentals of Optics*, McGraw-Hill, New York, 1957). They are defined by

$$C(v) = \int_0^v \cos \frac{\pi v^2}{2} dv \qquad (5.24)$$

and

$$S(v) = \int_0^v \sin \frac{\pi v^2}{2} dv \qquad (5.25)$$

Values of the Fresnel integrals in the range $0 < v < 1.0$ are presented in Table 5.2.

The excess path loss as given by Equation (5.22) is plotted as a function of the diffraction parameter in Figure 5.7. Several features of this figure deserve comment. Note that when the direct ray just grazes the top of the absorbing screen, making both h_e and v equal to zero, the excess path loss is still 6 dB. The excess path loss does not decrease to zero until the edge of the screen is withdrawn, so that it falls below the edge of the direct path by a significant amount given by

$$h_e = -0.6 \, r_1 \qquad (5.26)$$

TABLE 5.2 Values of Fresnel Integrals

Diffraction Parameter, v	Fresnel Cosine Integral, $C(v)$	Fresnel Sine Integral, $S(v)$
0.00	0.0000	0.0000
0.10	0.1000	0.0005
0.20	0.1999	0.0042
0.30	0.2994	0.0141
0.40	0.3975	0.0334
0.50	0.4923	0.0647
0.60	0.5811	0.1105
0.70	0.6597	0.1721
0.80	0.7230	0.2493
0.90	0.7648	0.3398
0.10	0.7799	0.4383

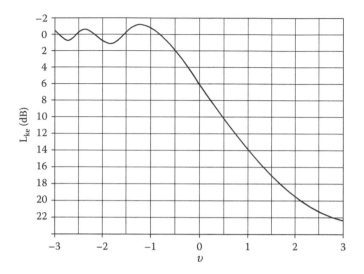

FIGURE 5.7 L_{ke} versus v for diffraction by an absorbing screen with a knife-edge.

where

$$r_1 = \sqrt{\frac{\lambda d_1 d_2}{(d_1 + d_2)}} \qquad (5.27)$$

The distance r_1 is called the radius of the first Fresnel zone, a region bounded by the ellipsoid that would be traced out if a string of length $d_1 + d_2 + \lambda/2$ were to be anchored at the transmitter and receiver points and stretched by a pencil tracing out the ellipsoid. This ellipsoid is pictured in Figure 5.8 where the sum of the distance from the transmitter, a, and the distance from the receiver, b, is given by

$$a + b = d_1 + d_2 + \lambda/2 \qquad (5.28)$$

Returning to the plot of the excess path loss in Figure 5.7, we see that as the top of the absorbing screen recedes from the direct path, so that $h_e < -0.6\ r_1$ (or when $v < -0.84$), the value of L_{ke}(dB) oscillates about 0 dB (i.e., for certain values of the screen top position below the direct path, the received signal power exceeds the power that would be received in the absence of the screen). On the other hand, when the screen is inserted so that $h_e > -0.6\ r_1$ and $v > -0.84$, excess path loss increases

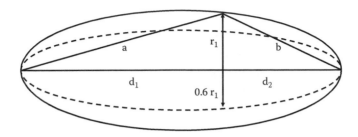

FIGURE 5.8 Ellipsoid enclosing the first Fresnel zone.

monotonically as h_e and v increase. When the screen top is well above the direct path and the diffraction parameter $v > 1$, to good approximation, the excess path loss may be evaluated from the following equation:

$$L_{ke} = 20\log(\pi v\sqrt{2}) \tag{5.29}$$

Sample Calculation:

Consider a 3 GHz transmitter separated from a receiver by a distance of 5 km. An earthen ridge with a straight, sharp top is located 2 km from the transmitter and juts 25 m above the direct path between the transmitter and receiver.

Estimate the path loss by modeling the ridge as an absorbing screen with a knife-edge top.

Solution:

The relevant dimensions are $d_1 = 2000$ m, $d_2 = 3000$ m, $h_e = 25$ m, and $\lambda = 0.1$ m.

Then from Equation (5.21), the diffraction parameter is

$$v = 25\ [(2 \times 5000)/(0.1 \times 2000 \times 3000)]^{1/2} = 3.23$$

The excess path loss is found from Equation (5.29) as

$$L_{ke}(dB) = 20 \log\ [3.14 \times 3.23 \times 1.41] = 23.1\ dB$$

The free space path loss is found from Equation (5.6) as

$$L_F\ (dB) = 32.4 + 20 \log 5 + 20 \log 3000 = 115.8\ dB$$

Finally, from Equation (5.20), the total path loss is
$$L\ (dB) = 115.8 + 23.1 = 138.9\ dB$$

5.2.2 Deygout Method of Approximately Treating Multiple Diffracting Edges

When there are multiple and sequential obstructions, the excess path loss they cause cannot simply be found by combining the individual losses that would be caused by each obstruction in the absence of the others. The fields reaching an obstruction are strongly influenced by the previous obstructions.

J. Deygout proposed a method of estimating the excess path loss due to multiple obstructions that will be described now with reference to Figure 5.9, where, as an example, three absorbing screens block the direct path between transmitter and receiver. First, the diffraction parameter is calculated for each edge as if it were present alone. These single-edge diffraction parameters are as follows:

For knife-edge screen #1, $v_1 = v(d_1, d_2 + d_3 + d_4, h_1)$ (5.30a)

For knife-edge screen #2, $v_2 = v(d_1 + d_2, d_3 + d_4, h_2)$ (5.30b)

For knife-edge screen #3, $v_3 = v(d_1 + d_2 + d_3, d_4, h_3)$ (5.30c)

In Equation (5.30), the first term in parentheses is the distance from the transmitter to the designated screen, the second term is the distance from the screen to the receiver, and the third term is the excess height of the designated screen with respect to the direct path between transmitter and receiver.

Comparing the three values of v in Equation (5.30), we choose the largest value and designate the corresponding screen

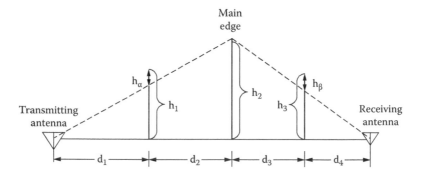

FIGURE 5.9 Three absorbing screens with knife edges blocking the direct wave path (Deygout method).

knife-edge as the main edge. For the example shown in Figure 5.9, edge #2 is the main edge. Then, subpaths are drawn. Subpath #1 is from the transmitter to the top of the main edge, and subpath #2 is from the main edge to the receiver. If now only one screen blocks each subpath as shown, one calculates values of the diffraction parameters for the subpaths:

$$v_1' = v(d_1, d_2, h_\alpha) \tag{5.31a}$$

where h_α is the excess height of edge #1 with respect to subpath #1, and

$$v_3' = v(d_3, d_4, h_\beta) \tag{5.31b}$$

where h_β is the excess height of edge #3 with respect to subpath #2. The total excess diffraction loss is then given by

$$L_{ex} = L_{ke}(v_1') + L_{ke}(v_2) + L_{ke}(v_3') \tag{5.32}$$

For the more general case, where there is more than one screen blocking a subpath, one must determine the main peak as before, and divide the subpath into sub-subpaths.

5.2.3 The Causebrook Correction to the Deygout Method

The Deygout method will tend to overestimate the excess path loss caused by multiple screens. This may be seen by considering the situation shown in Figure 5.10, where the three screens are inserted so that they just touch the direct path between transmitter and receiver (i.e., their excess heights are zero). Thus, we see that if we use Deygout's method,

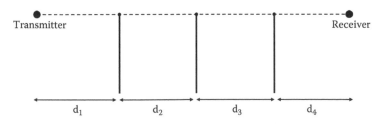

FIGURE 5.10 Wave path with three absorbing screens whose tops just reach the direct path (i.e., for each screen, the excess height, $h_e = 0$).

$$L_{ke}(v_1') = L_{ke}(v_2') = L_{ke}(v_3') = L_{ke}(0) = 6dB$$

and

$$L_{ex} = 18dB$$

where we have used Equation (5.32).

This value of excess path loss is clearly too large. The screen closest to the transmitter will absorb all the power in the original wave that falls below the direct path between transmitter and receiver, and cause a 6 dB loss in the received signal. However, the second screen is in the shadow region of the first screen and will have a far smaller value of power density falling in it, compared with the first screen; it will consequently absorb less power and will cause a signal loss smaller than 6 dB.

J.H. Causebrook proposed that correction factors be subtracted from the value of excess path loss calculated by the Deygout method to obtain a corrected value for excess path loss as given by

$$L_{ex}^{corr} = L_1' + L_2 + L_3' - C_1 - C_2 \tag{5.33}$$

C_1 and C_2 are the Causebrook correction factors defined by

$$C_1 = (6 - L_2 + L_1)\cos\alpha_1 \tag{5.34a}$$

where

$$\cos\alpha_1 = \sqrt{\frac{d_1(d_3 + d_4)}{(d_1 + d_2)(d_2 + d_3 + d_4)}}$$

and

$$C_2 = (6 - L_2 + L_3)\cos\alpha_3 \tag{5.34b}$$

where

$$\cos\alpha_3 = \sqrt{\frac{(d_1 + d_2)d_4}{(d_1 + d_2 + d_3)(d_3 + d_4)}}$$

In Equation (5.34), L_1, L_2, and L_3 are the values calculated as if each edge existed on its own.

We can now calculate these correction factors for the case of the three edges with zero excess height shown in Figure 5.10, and with $d_1 = d_2 = d_3 = d_4$.

From Equation (5.34),

$$C_1 = (6 - L_2 + L_1)\sqrt{\frac{d_1(d_3 + d_4)}{(d_1 + d_2)(d_2 + d_3 + d_4)}} = 6\sqrt{\frac{1 \times 2}{2 \times 3}} = 3.5 \text{ dB}$$

Similarly, $C_2 = 3.5$ dB.

Then, the corrected value of excess path loss can be found from Equation (5.29) as

$$L_{ex}^{corr} = 18dB - C_1 - C_2 = 11dB$$

In Figure 5.11, both the values of excess path loss obtained by the Deygout method without correction, and those obtained by applying the Causebrook correction are plotted as a function of the number of equally spaced edges, whose tops just touch the direct path between transmitter and receiver. Also plotted are the values of excess path loss obtained by L.E. Volger. Volger calculated the fields

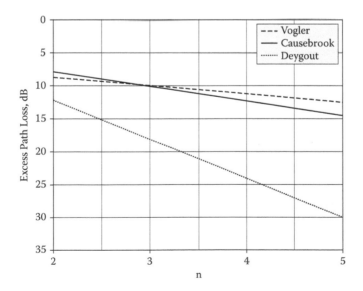

FIGURE 5.11 Excess path loss for n screens with $h_e = 0$.

after each screen and applied them to the succeeding screen. His results may be regarded as being equally as valid for multiple screens as the Huygens–Fresnel results for a single diffracting knife-edge screen. Thus, we see from the figure that the Deygout method with the Causebrook correction gives reasonably accurate values of excess path loss at least when only a few edges are involved.

5.3 Wave Propagation in an Urban Environment

5.3.1 The Delisle/Egli Empirical Expression for Path Loss

G.Y. Delisle has proposed an empirical formula for estimating path loss in densely built-up urban areas; the formula is based on the method of J.J. Egli, which averages a large number of measurements in U.S. cities. The empirical formula takes the following form:

$$L_{empirical}(dB) = 40 \ \log r_{km} + 20 \ \log f_{MHz} - 20 \ \log \underline{h}_b$$
$$+76.3 - 10 \ \log \underline{h}_m \tag{5.35a}$$

for $h_m < 10$ m, and

$$L_{empirical}(dB) = 40 \ \log r_{km} + 20 \ \log f_{MHz} - 20$$
$$\log \underline{h}_b + 86.3 - 20 \ \log \underline{h}_m \tag{5.35b}$$

for $h_m > 10$ m, where r_{km} is the range normalized to 1 km, f_{MHz} is the frequency normalized to 1 MHz, h_b is the height of the base station antenna normalized to 1 m, and h_m is the height of the mobile antenna normalized to 1 m. Either Equation (5.35a) or Equation (5.35b) may be used when $h_m = 10$ m.

Converting from decibel notation to ordinary numbers, Equation (5.35) becomes

$$L_{empirical} = 4.27 \times 10^{-17} \frac{r^4 f^2}{h_b^2 h_m} \tag{5.36a}$$

for $h_m < 10$ m, and

$$L_{empirical} = 4.27 \times 10^{-16} \frac{r^4 f^2}{h_b^2 h_m^2} \qquad (5.36b)$$

for $h_m > 10$ m.

The parametric dependences of path loss in Equation (5.36) are certainly at variance with the predictions of the free space propagation model, which predicts range dependence for the path loss as r^2 and has no dependence on antenna heights. It is also at variance with the plane earth propagation model, which has no dependence on signal frequency.

In the next few subsections, we will describe a physical model of wave propagation in urban areas that is in reasonable agreement with the empirical Equation (5.36). As anticipated, it will involve diffraction by multiple obstructions (tops of buildings), plus reflection and phase interference in the final street. The situation to be modeled is shown in Figure 5.12.

5.3.2 The Flat-Edge Model for Path Loss from the Base Station to the Final Street

It is tractable but complicated to numerically evaluate the electromagnetic field produced by a series of absorbing screens with knife edges. The field produced by the first screen in the series is evaluated and applied to the second screen. The resulting field produced by the second screen is evaluated and applied to the third screen, and so on. The flat-edge model, due to S.R. Saunders and F.R. Bonar, simplifies the analysis by assuming that all the screens are equally spaced and of the same height. It has been applied to account for excess diffraction loss due to the buildings along a direct path from the base station antenna to the far edge of the building at the start of the final street. It models each building as an absorbing screen with a knife edge. The model is

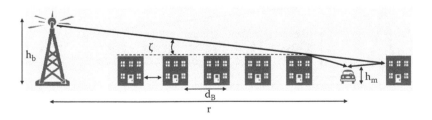

FIGURE 5.12 Typical propagation path in an urban area.

shown in Figure 5.13. The edges have negative excess height, but as was seen in Section 5.2.2, even such edges can contribute to excess path loss.

Important quantities in this model are as follows:

n_B, the number of buildings (edges) that are close enough to the direct path to cause significant diffraction

d_B, the spacing between adjacent edges

r_B, the distance from the base station to the first diffracting edge

h_o, the height of the buildings (edges)

h_b, the height of the base station antenna

ζ, the angle between the direct path and a horizontal line across the tops of the buildings

L_n, the excess path loss due to diffraction by the n_B building edges

The flat-edge model assumes that $r_B \gg n_B d_B$. An important parameter in the analysis is

$$\kappa = -\zeta(\pi d_B/\lambda)^{1/2} \qquad (5.37)$$

In Figure 5.14, results of analyzing the flat-edge model are presented as a plot of excess path loss versus the number of edges with κ as a parameter. For $\kappa = 0$, the direct path just grazes the tops of the edges, and the result is the same as given by Vogler's analysis as plotted in Figure 5.11.

The curves for larger negative values of κ in Figure 5.14 have a number of interesting features. (κ becomes increasingly negative as the height of the base station antenna rises above the height of the buildings.) In general, for a small number of edges, these curves have regions where excess path loss decreases as the number of edges is increased; this reflects a phenomenon shown in Figure 5.7, where it is seen that an edge with certain values of

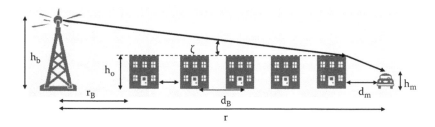

FIGURE 5.13 Flat-edge model for propagation from base station antenna to edge of building at start of final street. Each building is modeled as an absorbing screen with knife-edge.

negative excess height with $h_e < -0.6\ r_1$ can actually cause a reduction in path loss.

One can use Figure 5.14 to estimate the excess path loss due to diffraction by the building edges with $1 < n_B < 100$ and $-1 < \kappa < 0$. Alternatively, one can estimate the excess path loss using the approximate expression

$$L_n\ (\text{dB}) = -(3.29 + 9.9 \log n_B) \log (-\kappa) - (0.77 + 0.26 \log n_B) \tag{5.38}$$

For $\kappa = -0.1$ and $n_B = 10$, Equation (5.38) gives an excess path loss of 12.2 dB in agreement with the plot in Figure 5.14. However, for $\kappa = -0.1$ and $n_B = 100$, Equation (5.38) gives an excess path loss of 21.8 dB, while Figure 5.14 indicates that the correct value of excess path loss is only 15 dB. Thus, the reader is cautioned that Equation (5.38) may only give reasonably accurate values when the number of buildings is relatively small.

5.3.3 Ikegami Model of Excess Path Loss in the Final Street

After diffraction by the edge of the final building, the signal may reach the mobile antenna by two paths:

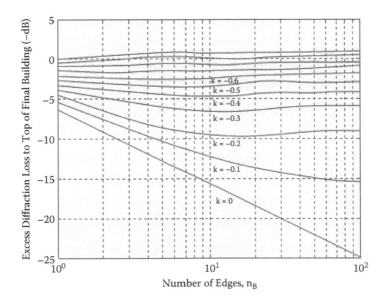

FIGURE 5.14 Results of flat-edge model analysis with $d_B = 150\lambda$.

1. Directly from the final edge to the antenna
2. From the final edge to the wall of the opposite building, and then to the antenna by specular reflection

The waves on the two paths of different lengths will have different phases when they arrive at the antenna, and this phase difference must be taken into account when summing the two field contributions. The situation is somewhat reminiscent of the plane earth model described in Section 5.1.4.

The final street and the two wave paths are shown in Figure 5.15. This geometry has been analyzed by F. Ikegami with appropriate assumptions concerning the reflectivity of the opposite wall. His result for excess path loss in the final street is given by

$$L_I(dB) = 10 \ \log f_{MHz} + 20 \ \log(\underline{h}_0 - \underline{h}_m) - 10 \ \log \underline{d}_s - 22.7 \ (5.39)$$

where h_o, h_m, and d_s are all normalized to 1 m.

The total path loss for the urban propagation situation in Figure 5.12, is the sum of the free space path loss, $L_F(dB)$, the excess path loss due to diffraction by the building edges, $L_n(dB)$, and the excess path loss due to reflection and towpath phase interference in the final street, $L_I(dB)$; that is,

$$L(dB) = L_F(dB) + L_n(dB) + L_I(dB) \qquad (5.40)$$

5.3.4 The Walfisch–Bertoni Analysis of the Parametric Dependence of Path Loss

For large values of n_B in Figure 5.14, the values of excess path loss tend to be constant as more edges are added. The value of n_B at

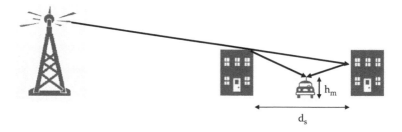

FIGURE 5.15 Ikegami model of propagation in the final street.

which this phenomenon begins is known as the settled value, n_s, given by

$$n_s = \pi/\kappa^2 \tag{5.41}$$

In the settled region, $n_B > n_s$, the excess path loss is a function of κ but is independent of the number of edges.

J. Walfisch and H.L. Bertoni had previously noted the tendency of excess path loss caused by diffraction over building tops to become independent of the number of buildings in the settled field region. They pointed out specifically that for a significant range of κ (viz., $0.03 < \kappa < 0.4$),

$$L_n \sim (-\kappa)^{-1.8} \tag{5.42}$$

where we recall from Equation (5.37) that

$$-\kappa \sim \zeta\, f^{1/2} \tag{5.43}$$

and

$$\zeta = \tan^{-1}[(h_b - h_o)/r] \tag{5.44}$$

When the following inequalities pertain,

$$r \gg h_b \gg h_o$$

Equation (5.44) may be replaced by

$$\zeta \approx h_b/r \tag{5.45}$$

Then, using Equations (5.45) and (5.43), Equation (5.42) gives for the parametric dependence of the diffraction excess path loss

$$L_n \sim h_b^{-1.8}\, r^{1.8}\, f^{-0.9} \tag{5.46}$$

The free space path loss

$$L_F \sim r^2 f^2 \tag{5.47}$$

The parametric dependence of the final street path loss may be obtained from Equation (5.39) as

$$L_I \sim f(h_o - h_m)^2 \qquad (5.48)$$

Then, by combining Equations (5.46), (5.47), and (5.48), we may obtain the parametric dependencies of the total path loss as

$$L = L_F L_n L_I \sim r^{3.8} f^{2.1} h_b^{-1.8} (h_o - h_m)^2 \qquad (5.49)$$

Parametric dependencies of various models are compared with empirical results in Table 5.3.

It is clear from Table 5.3 that the free space and plane earth models have major discrepancies when compared with empirical results in urban areas. On the other hand, the Walfisch–Bertoni limit of the flat-edge diffraction model, combined with the Ikegami model of reflection and phase interference in the final street, produce agreement with empirically observed average path loss dependence on range, frequency, and base station antenna height.

Even the dependence on mobile antenna height agrees somewhat with the empirical expressions. In Figure 5.16, we plotted $1 - (h_m/h_o)^2$ versus (h_m/h_o). In the interval $0.5 < (h_m/h_o) < 0.64$, the slope approximately matches an h_m^{-1} dependence, while for $0.64 < (h_m/h_o) < 0.8$, the slope approximately matches an h_m^{-2} dependence. Note that for an average building height of 16 m, $(h_m/h_o) = 0.64$ corresponds to $h_m = 10$ m, in agreement with the empirical transition point as noted in Table 5.3.

TABLE 5.3 Comparison of Parametric Dependences of Path Loss Models with Empirical Observations

Model	r Dependence	f Dependence	h_b Dependence	h_m Dependence
Egli (empirical)	r^4	f^2	h_b^{-2}	h_m^{-1}, $(h_m < 10$ m) h_m^{-2}, $(h_m > 10$ m)
Free space	r^2	f^2	No dependence	No dependence
Plane earth	r^4	No dependence	h_b^{-2}	h_m^{-2}
Walfisch–Bertoni + Ikegami	$r^{3.8}$	$f^{2.1}$	$h_b^{-1.8}$	$(h_o - h_m)^2 = h_o^2$ $[1 - (h_m/h_o)^2]$

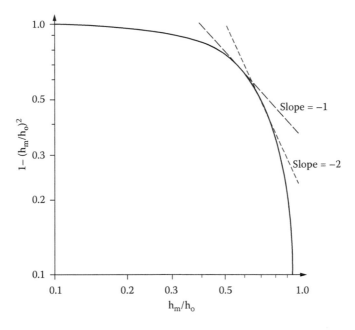

FIGURE 5.16 $[1 - (h_m/h_o)^2]$ vs h_m/h_o.

Problems

5.1. A right-hand circularly polarized plane wave is incident on the boundary between dry air and dry earth. Describe the polarization of the reflected wave for the following cases:
   a. Angle of incidence smaller than the Brewster angle
   b. Angle of incidence equal to the Brewster angle
   c. Angle of incidence greater than the Brewster angle

5.2. A linearly polarized plane wave with power density $S = 5 \times 10^{-9}$ Watts per square meter is incident from free space onto the flat surface of a ferrite having relative permeability of 100 and relative permittivity of 4. The angle of incidence is 45°, and the incident wave is polarized parallel to the plane of incidence.
   a. Calculate the angle of transmission.
   b. Calculate the electric field strength in the transmitted wave.
   c. Calculate the electric field strength in the reflected wave.

d. Check that your answers are compatible with the following relationship between the reflection and transmission coefficients:

$$1 + R_\parallel = T_\parallel \cos \theta_t / \cos \theta_i$$

5.3. If the standard deviation of the probability density function for the height of a rough surface is 50 cm, estimate the range of the angle of incidence for an incoming plane wave at a frequency of 900 MHz at which the surface will appear to be approximately flat.

5.4. A communication system operating with a center frequency of 900 MHz has a 20 Watt transmitter feeding a 21.6 dBi antenna array on top of a 50 m high tower through a cable with 5 dB loss. The receiving antenna is handheld at a height of 1.5 m, has a gain of 2.16 dBi, and feeds a matched receiver through a cable with 1 dB loss. To be understandable in the presence of noise, the minimum acceptable received signal power is −110 dBm.

a. Calculate the maximum range using free space path loss.

b. Calculate the maximum range using plane earth path loss.

c. Calculate the maximum range using Egli empirical path loss.

5.5. A 1.8 GHz transmitter and a receiver are separated by 6 km. A sharp edge absorbing screen is inserted below and normal to the direct path between transmitter and receiver, which is 2 km from the transmitter. How closely can the edge of the screen approach the direct path and not introduce excess path loss?

5.6. If in the communication system described in Problem 5.5, the screen is inserted so that its edge is 20 m above the direct path, calculate the excess path loss and the total path loss.

5.7. A 1 GHz transmitter sends a signal to a receiver 4 km away. There are three knife-edged absorbing screens intercepting the direct path. The first screen has an excess height of 5 m and is 1 km from the transmitter. The second screen has an excess height of 10 m and is 2 km from the transmitter. The third screen has an excess height of 5 m and is 3 km from the transmitter. Find the excess path loss using the Deygout method.

5.8. Repeat Problem 5.7 using the Causebrook correction to find a better value of the excess path loss.

5.9. Use the empirical path loss formula to calculate the range for a maximum acceptable path loss of 115 dB, given that the height of the base station antenna is 50 m and the height of the mobile antenna is 1.5 m.

   a. When the frequency is 900 MHz
   b. When the frequency is 1800 MHz

5.10. For a propagation range of R = 4.15 km, base station antenna height of h_b = 100 m, building height of h_o = 15 m, building spacing of d_B = 40 m, and f = 900 MHz, calculate the flat-edge model γ parameter, and n_s, the number of buildings required for the field to settle. Assuming that the number of buildings, $n_B = n_s$, discuss the validity of the Walfisch–Bertoni analysis.

5.11. Use the flat-edge model plus the Ikegami model to calculate the total path loss given that f = 900 MHz, r = 2.9 km, h_b = 30 m, h_o = 15 m, h_m = 1.5 m, d_B = 40 m, d_s = 20 m, and n_B = 10.

5.12. For the parameters given in Problem 5.11 calculate the path loss both by using the empirical expression and by using the plane earth model. Comment on the degree of agreement obtained between the empirical expression and the model used in Problem 5.11.

Bibliography

1. J. Walfisch and H.L. Bertoni, "A Theoretical Model of UHF Propagation in Urban Environments," *IEEE Transactions on Antennas and Propagation* 36 (1988): 1788–1796.

2. L.E. Volger, "The Attenuation of Electromagnetic Waves by Multiple Knife-Edge Diffraction," U.S. Dept. of Commerce, NTIA Report 81-86 (1981).

3. F. Ikegami, T. Takeuchi, and S. Yoshida, "Theoretical Predictions of Mean Field Strength for Urban Mobile Radio," *IEEE Transactions on Antennas and Propagation* 39 (1991): 299–302.

4. J.J. Egli, "Radio Propagation above 40 MC over Irregular Terrain," *Proceedings of IRE*, pp. 1383–1391 (1957).

5. S.R. Saunders and F.R. Bonar, "Explicit Multiple Building Diffraction Attenuation Function for Mobile Radio," *Electronic Letters* 27 (1991): 1276–1277.

6. G.Y. Delisle, J. Lefevre, M. Lecours, and J. Chouinard, "Propagation Loss Prediction: A Comparative Study with Application to the Mobile Radio Channel," *IEEE Transactions on Vehicular Technology* 26 (1985): 295–308.

7. J. Deygout, "Multiple Knife-Edge Diffraction of Microwaves," *IEEE Transactions on Antennas and Propagation* 14 (1966): 480–489.

8. S.R. Saunders, *Antennas and Propagation for Wireless Communication Systems* (John Wiley & Sons, Chichester, UK, 1999), 33–50, 111–118, 152–170.

9. F.A. Jenkins and H.E. White, *Fundamentals of Optics* (McGraw-Hill, New York, 1957).

10. M. Born and E. Wolf, *Principles of Optics*, 4th ed. (Pergamon Press, Oxford, UK, 1970).

11. A.H. Sihvola, *Electromagnetic Mixing Formulae and Applications* (Institution of Engineering and Technology, IET, London, 1999). See page 278 for permittivity of very dry earth.

12. International Telecommunications Union, ITU-R Recommendation 527-3, "Electrical Characteristics of the Surface of the Earth," Geneva, 1992.

CHAPTER **6**

Statistical Considerations In Designing Cell Phone Systems and Wireless Local Area Networks (WLANs)

The expressions for path loss discussed in Chapter 5 will yield values for average path loss. Actually, real values of path loss will be larger or smaller than the average value, depending on the mix of buildings on a particular path between the base station antenna and the mobile. If the design of a cell phone system only allows for the average value of path loss, the expected probability of a successful call would be only 50%. This is unsatisfactory, and one must learn how to compensate for the stochastic component of path loss. We begin by reviewing some fundamental concepts in statistical analysis.

6.1 A Brief Review of Statistical Analysis

6.1.1 Random Variables

A random variable is a function that, instead of assigning a unique numerical value to the outcome of an experiment, assigns a probability that the outcome will fall within a given range of values. For a continuous random variable, this range of values is continuous. As an example, the lifetime of the filament in an electric lightbulb is a continuous random variable with expected values falling in the range between zero and infinity.

In general, a random variable X has a probability distribution function (p.d.f.) $p(X)$ such that the probability of X having a value in the range from $X(1)$ to $X(2)$ is

$$\Pr[X(1) < X < X(2)] = \int_{X(1)}^{X(2)} p(X)dX \qquad (6.1)$$

Of course, because the random variable must assume some value, all probability density functions have the property that

$$\int_{-\infty}^{\infty} p(X)dX = 1 \qquad (6.2)$$

The expected value of X is called its mean value and is given by

$$\mathrm{EXP}[X] = \int_{-\infty}^{\infty} Xp(X)dX = \mu \qquad (6.3)$$

As indicated, the mean value is denoted by the symbol μ.

If one has a function of X denoted by $f(X)$, its expected value is calculated using the following equation:

$$\mathrm{EXP}[f(X)] = \int_{-\infty}^{\infty} f(X)p(X)dX \qquad (6.4)$$

The variance of X is given by

$$\sigma^2 = \mathrm{EXP}[(X - \mu)^2] = \mathrm{EXP}[X^2] - \mu^2 \qquad (6.5)$$

The square root of the variance, σ, is called the standard deviation.

Two random variables, X and Y, have a joint p.d.f., $p(X,Y)$, such that the probability of X having values between $X(1)$ and $X(2)$, and Y having values between $Y(3)$ and $Y(4)$ is given by

$$\Pr[X(1) < X < X(2), Y(3) < Y < Y(4)] = \int_{X(1)}^{X(2)} \int_{Y(3)}^{Y(4)} p(X,Y)dXdY \quad (6.6)$$

The cross-correlation between the random variables X and Y is given by

$$R_{XY} = EXP[XY^*] = \int_{-\infty}^{\infty}\int_{-\infty}^{\infty} XY^* p(X,Y)dXdY$$

The random variables are independent of each other if

$$p(X,Y) = p(X)\, p(Y) \tag{6.7}$$

6.1.2 Random Processes

$X(t)$ is a random process when its value at any given time $t = t_1$ [i.e., $X(t_1)$] is a random variable.

A random process is said to be stationary if its statistical properties do not change with time; that is,

$$p[X(t_1)] = p[X(t_1 + \tau)] \tag{6.8}$$

The autocorrelation function of $X(t)$ is defined by

$$R_{XX}(t_1, t_1 + \tau) = EXP[X(t_1)\, X^*(t_1 + \tau)] \tag{6.9}$$

A random process is said to be "wide-sense stationary" if R_{XX} only depends on the time shift τ and not on t_1; that is,

$$R_{XX}(t_1, t_1 + \tau) = R_{XX}(\tau) \tag{6.10}$$

Equation (6.10) is usually satisfied for random processes encountered in describing communication systems.

6.2 Shadowing

Path loss will have a mean value $\bar{L}$ and a stochastic part L_s; that is,

$$L(dB) = \bar{L} + L_s \tag{6.11}$$

L_s is called shadowing and is a random variable.

The shadowing will have zero mean, because any contribution to the mean value of path loss is absorbed in $\bar{L}$.

6.2.1 The Log-Normal Probability Distribution Function

The probability distribution function for the shadowing is a log-normal distribution function with zero mean of the form

$$p(L_s) = \frac{1}{\sigma_L \sqrt{2\pi}} e^{-\frac{L_s^2}{2\sigma_L^2}} \qquad (6.12)$$

where L_s is expressed in dB, and σ_L is the standard deviation of the shadowing.

Equation (6.12) is plotted in Figure 6.1.

6.2.2 The Complementary Cumulative Normal Distribution Function (Q Function)

The integral over the shaded region under the curve in Figure 6.1 represents the probability that the shadowing is greater than $2\sigma_L$. Such an integral is called the complementary cumulative normal distribution function and in this case is denoted by the symbol $Q(2)$, where the number 2 indicates the lower bound on the integration is $2\sigma_L$. In general, the Q function gives the probability that L_s is greater than a value M as

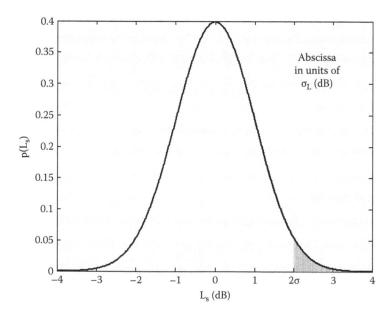

FIGURE 6.1 PDF of a random variable, L_s, with log-normal distribution.

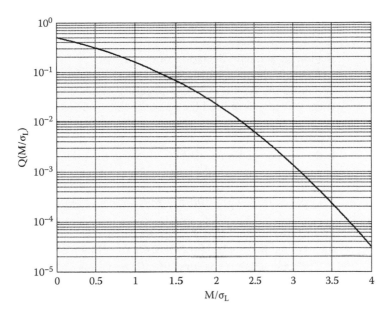

FIGURE 6.2 The Q function.

$$\Pr\left[L_S > M\right] = Q(M/\sigma_L) = \int\limits_{M}^{\infty} p(L_s)dL_s \qquad (6.13)$$

The Q function is plotted in Figure 6.2, and its values are tabulated in Table 6.1.

6.2.3 Calculating Margin and Probability of Call Completion

In designing a cell phone system we are concerned about the probability of being unable to complete a call successfully. This failure is called an outage, and its probability is denoted by p_{out}. As mentioned previously, if a cell phone system is designed allowing for only the average path loss, $p_{out} = 0.5$. To make p_{out} smaller, we must design the cell phone system with allowance for the possibility that path loss will be larger than its average value— that is, in our system design, we must allow for a path loss:

$$L(dB) = \bar{L} + M \qquad (6.14)$$

In Equation (6.14), M is called the margin.

If we allow for margin, then the probability of an outage becomes

$$p_{out} = \Pr\left[L_s > M\right] = Q(M/\sigma_L) \qquad (6.15)$$

Sample Problem

A cell phone system has a mean path loss of 100 dB and shadowing standard deviation of $\sigma_L = 7$ dB.

What is the total path loss that must be allowed for in designing the system if 98% probability of call completion at the cell edge is desired (i.e., $p_e = 98\%$)?

Answer

If the call completion probability is 98%, the probability of an outage is

$$p_{out} = 100\% - p_e = 100\% - 98\% = 2\% = Q(M/\sigma_L)$$

From Table 6.1, for $Q(M/\sigma_L) = 0.02$, $M/\sigma_L = 2.05374$. Then, the required margin is $M = 2.05374 \times \sigma_L = 14.4$ dB.

The total path loss that must be allowed for is $L(dB) = \overline{L} + M = 114.4$ dB.

6.2.4 Probability of Call Completion Averaged over a Cell

We have been considering the probability of call completion at the cell edge. Often what is specified is the probability of call

TABLE 6.1 Values of the Q(M/σ) Function

Q(M/σ)	M/σ	Q(M/σ)	M/σ	Q(M/σ)	M/σ	Q(M/σ)	M/σ
0.00	∞	0.05	1.64485	0.24	0.70630	0.43	0.1763
0.00003	4.00	0.06	1.55477	0.25	0.67449	0.44	0.1509
0.00009	3.75	0.07	1.47579	0.26	0.64335	0.45	0.12566
0.00023	3.50	0.08	1.40507	0.27	0.61281	0.46	0.10043
0.00058	3.25	0.09	1.34076	0.28	0.58284	0.47	0.07527
0.001	3.09023	0.10	1.28155	0.29	0.55338	0.48	0.05015
0.002	2.87814	0.11	1.22653	0.30	0.52440	0.49	0.02507
0.003	2.74778	0.12	1.17499	0.31	0.49585	0.50	0
0.004	2.65207	0.13	1.12639	0.32	0.46770		
0.005	2.52583	0.14	1.08032	0.33	0.43991		
0.006	2.51214	0.15	1.03643	0.34	0.41246		
0.008	2.40892	0.16	0.99446	0.35	0.38532		
0.01	2.32635	0.17	0.95416	0.36	0.35846		
0.015	2.17009	0.18	0.91537	0.37	0.33285		
0.02	2.05375	0.19	0.87790	0.38	0.30548		
0.025	1.95996	0.20	0.84162	0.39	0.27932		
0.03	1.88079	0.21	0.80642	0.40	0.25335		
0.035	1.81191	0.22	0.77219	0.41	0.22754		
0.04	1.75069	0.23	0.73885	0.42	0.20189		

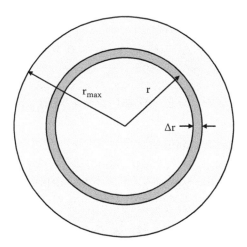

FIGURE 6.3   Cell incremental ring geometry.

completion averaged over the entire cell, p_{cell}. The relationship between p_{cell} and the probability of call completion to a mobile transceiver located at the outer edge of the cell, p_e, may be derived with reference to Figure 6.3.

At a radius $r < r_{max}$ from the base station at the center of the cell, where r_{max} is the distance from the base station to the edge of the cell, the probability of a successful call, $p_s(r)$, is larger than p_e. Specifically, if the maximum acceptable path loss is L_{max}, the margin $M(r) = L_{max} - \bar{L}(r)$, and

$$p_s(r) = 1 - Q(M/\sigma_L) = 1 - Q([L_{max} - \bar{L}(r)]/\sigma_L) \quad (6.16)$$

Then, averaging $p_s(r)$ over the cell we obtain

$$P_{cell} = \frac{1}{\pi r^2} \int_0^{r_{max}} p_s(r) 2\pi r dr = \frac{2}{r^2} \int_0^{r_{max}} \{1 - Q[\frac{L_{max} - \bar{L}}{\sigma_L}]\} r dr \quad (6.17)$$

The integral in Equation 6.17 can be evaluated analytically if the mean path loss has a power law dependence on range (i.e., $\bar{L} \sim r^n$). In that case, the expression for probability of call completion averaged over the cell is

$$P_{cell} = p_e + \exp(A') \, Q(B') \quad (6.18)$$

where

$$A' = 0.106 \, \sigma_L^2/n^2 + 0.46 \, M/n$$

and

$$B' = 0.461 \, \sigma_L/n + M/\sigma_L$$

Sample Calculation:

The probability of call completion at the cell edge is $p_e = 0.9$. The exponent for path loss dependence on range is $n = 4$. The standard deviation of the shadowing is $\sigma_L = 16$ dB.

Use Equation (6.18) to find the probability of call completion averaged over the cell.

Solution:

First, calculate the margin, M:

$$Q(M/\sigma_L) = 1 - p_e = 0.1$$

Then, from Table 6.1,

$$M/\sigma_L = 1.28$$

and

$$M = 1.28 \times 16 \text{ dB} = 20.5 \text{ dB}$$

Thus, the parameter A' may be evaluated as

$$A' = 0.106 \times (16/4)^2 + 0.46 \times (20.5/4) = 4.06$$

and the parameter B' may be evaluated as

$$B' = 0.461 \times (16/4) + 1.28 = 3.12$$

Then, the probability of call completion averaged over the cell is obtained from Equation (6.18) as

$$p_{cell} = 0.9 + \exp(4.06) \, Q(3.12) = 0.9 + 57.7 \times 0.00092 = 0.95$$

The value of $Q(3.12)$ was found by interpolating from the values in Table 6.1.

Thus, it is seen that in this case, where the probability of call completion at the cell edge is 90%, the probability of call completion averaged over the cell is 95%.

6.2.5 Additional Signal Loss from Propagating into Buildings

Usually, it is desirable to have cell phones operate inside buildings. The additional signal attenuation in passing through building walls can be considerable, especially in urban areas where building walls may be concrete reinforced with iron bars. In Figure 6.4, typical values of mean signal loss encountered in penetrating a building are plotted versus frequency. Three cases are shown: residential buildings, urban buildings, and an intermediate case.

There will be variation in building loss depending on the structural details of individual buildings. The building loss will have a random part, L_{BS}, in addition to the mean part, $\overline{L}_B$; that is,

$$L_B = \overline{L}_B + L_{BS} \tag{6.19}$$

L_{BS} is a random variable with zero mean and a log normal probability distribution.

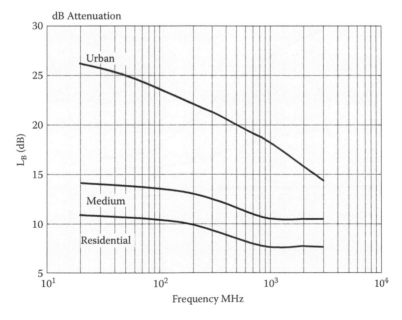

FIGURE 6.4 Mean building attenuation.

Typical values of the standard deviation of L_{BS} are plotted in Figure 6.5 as a function of frequency. Again, three cases are shown: residential buildings, urban buildings, and the intermediate case.

The question now arises as to how one combines path loss with building loss. In fact, the process is rather straightforward, because the stochastic part of each loss follows log-normal statistics. In that case, the mean values add, and the standard deviations add as the square root of the sum of the squares; the combined mean loss is

$$\overline{L}_{total} = \overline{L}_{path\ loss} + \overline{L}_B \qquad (6.20)$$

while the total standard variation is

$$\sigma_{total} = (\sigma_L^2 + \sigma_B^2)^{1/2} \qquad (6.21)$$

6.2.6 Shadowing Autocorrelation (Serial Correlation)

As a cell phone moves, the mix of obstructions between it and the base station changes. In Figure 6.6, this change in path obstructions is represented by the differing shadowing losses: L_{11} for cell phone position #1 and L_{12} for cell phone position #2. The shadowing losses L_{11} and L_{12} are each zero mean log-normal random variables;

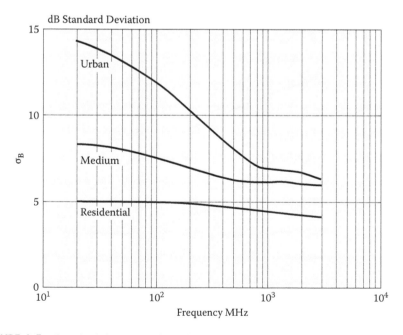

FIGURE 6.5 Standard deviation of building attenuation.

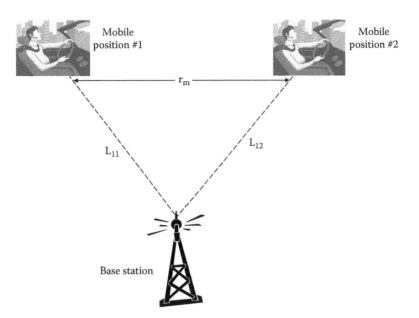

FIGURE 6.6   Geometry for shadowing serial correlation.

however, they are not independent of each other because the two paths may contain some of the same obstructions, especially when r_m, the distance between positions #1 and #2 is small.

We can define an autocorrelation or serial correlation function as

$$\rho_s(r_m) = EXP[L_{11}L_{12}]/(\sigma_1\sigma_2) \approx EXP[L_{11}L_{12}]/\sigma_L^2 \quad (6.22)$$

where the expected value of $L_{11}L_{12}$ is given by

$$EXP[L_{11}L_{12}] = \iint L_{11}L_{12}\, p(L_{11})\, p(L_{12})\, d\,L_{11}\, d\,L_{12} \quad (6.23)$$

and where the probability density function of the shadowing loss is

$$p(L_s) = \exp(-L_s^2/2\,\sigma_L^2)/[(2\pi)^{1/2}\,\sigma_L] \quad (6.24)$$

As a check, we note that when $r_m \to \infty$, L_{11} and L_{12} should be uncorrelated, and Equation (6.23) becomes

$$\lim (r_m \to \infty)\, \{EXP[L_{11}L_{12}]\} = \int L_{11}\, p(L_{11})\, d\,L_{11}\, d\,L_{12} \int L_{1\,2}\, p(L_{12})\, d\,L_{12}$$

$$= \overline{L}_{11}\,\overline{L}_{12} = 0$$

On the other hand, if $r_m \to 0$, $L_{11} = L_{12} = L_s$ and one would expect complete correlation (i.e., $\rho_s (r_m) = 1$). To confirm this we calculate from Equation (6.23) that $\lim(r_m \to 0)\{EXP[L_{11}L_{12}]\} = \int L_s^2 \, p(L_s) \, d \, L_s \int p(L_s) \, d \, L_s = \sigma_L^2$
and then from Equation (6.22),

$$\lim(r_m \to 0)\{\rho_s (r_m)\} = \sigma_L^2/\sigma_L^2 = 1$$

In general ρ_s has a decaying exponential dependence on r_m. The value of r_m for which $\rho_s = e^{-1}$ is designated as $r_m = r_e$; r_e is called the shadowing correlation distance. This distance gives the scale length over which variation in shadowing is acceptably small. For $r_m > r_e$, power levels have to be changed. The shadowing correlation distance increases with distance from the base station; typical values are $r_e = 44$ m at 1.6 km from the base station, and $r_e = 112$ m at 4.8 km from the base station.

6.2.7 Shadowing Cross-Correlation

As discussed in Chapter 1, co-channel interference from an out-of-cell base station is kept to a tolerable level by ensuring that the distance from the mobile to the in-cell base station is much smaller than the distance from the mobile to the interfering base station.

Considering only interference from one out-of-cell base station, the ratio of in-cell signal power to interfering signal power received at the mobile is

$$P_s/P_I = 40 \log (d_i/r_c) \tag{6.25}$$

where d_i is the distance from the mobile to the interfering base station, and r_c is the cell radius.

However, when shadowing is present along the two paths, P_s/P_I becomes a random variable that we shall designate by the symbol Ω. One becomes concerned that the shadowing along the path to the in-cell base station may be greater than along the path to the interfering base station. For example, the path to the in-cell base station may involve diffraction over tall buildings, while the path to the interfering base station may be across open parkland. Obviously, if shadowing along the two paths is strongly correlated, there is little cause for concern, but if the shadowing along the two paths is weakly correlated, the value of Ω may be smaller than its mean value, which is given by Equation (6.25), by an unacceptable amount.

Refering to Figure 6.7, the variance of Ω is given by

$$\sigma_\Omega{}^2 = \text{EXP}\,[\Omega^2] - \{\text{EXP}\,[\Omega]\}^2 = \sigma_1{}^2 + \sigma_2{}^2 - 2\,R_{12}\,\sigma_1\,\sigma_2 \quad (6.26)$$

where σ_1 is the standard deviation of the shadowing along path L_{11}, σ_2 is the standard deviation of the shadowing along path L_{21}, and R_{12} is the cross-correlation of the shadowing along paths L_{11} and L_{21}.

If the standard deviation of the shadowing is the same along the two paths—that is, if

$$\sigma_1 = \sigma_2 = \sigma_L \quad\quad\quad (6.27)$$

Then, Equation (6.26) reduces to

$$\sigma_\Omega{}^2 = 2\,\sigma_L{}^2\,(1 - R_{12}) \quad\quad\quad (6.28)$$

It is clear from Equation (6.28) that if shadowing along the two paths is completely correlated (i.e., if $R_{12} = 1$), then the standard deviation of the signal-to-interference ratio is zero, and Equation (6.25) for μ_Ω, the mean value of Ω, is sufficient for determining system performance. However, measurements of R_{12} by Graziano ("Propagation Correlations at 900 MHz," *IEEE Transactions on Vehicular Technology* 27 (1978): 182–189) indicate that values fall from a maximum of 0.7 for an angle between the two paths of zero degrees, decreasing to about 0.35 as the angle increases to 180°. If $R_{12} < 1$, the Ω ratio will be smaller than the mean value with a probability of 50%. If a greater

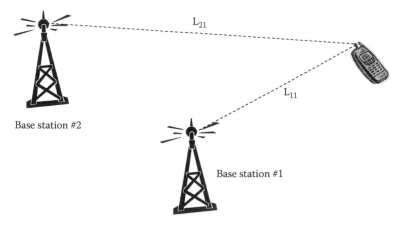

FIGURE 6.7 Geometry for shadowing cross-correlation.

probability of adequate Ω is required, the system can be redesigned say by increasing the frequency reuse distance, d_i, which effectively adds a margin, M_Ω, to the initial value of μ_Ω. The probability of adequate signal-to-interference ratio is then given by

$$\Pr\left[\Omega < \Omega_T\right] = 1 - Q\left(M_\Omega/\sigma_\Omega\right) \qquad (6.29)$$

where σ_Ω is given by Equation (6.28) and $\Omega_T = \mu_\Omega + M_\Omega$.

Sample Problem:

In an urban area, satisfactory cell phone service requires a signal-to-interference ratio, Ω, of at least 15 dB with a probability of 95%. Calculate the threshold value of the signal-to-interference ratio, Ω_T, for which a cell phone system must be designed. The standard deviation of the shadowing is $\sigma_L = 8$ dB, and the cross-correlation of shadowing along the in-cell path and the interfering path is estimated to be $R_{12} = 0.6$.

Solution:

From Equation (6.29) we require that

$$1 - Q\left(M_\Omega / \sigma_\Omega\right) = 0.95$$

or

$$Q\left(M_\Omega/\sigma_\Omega\right) = 0.05$$

From a table of Q function values (Table 6.1), we then find that

$$M_\Omega = 1.64485\ \sigma_\Omega$$

We calculate the standard deviation of the signal-to-interference ratio using Equation (6.28) as

$$\sigma_\Omega = [2 \times 8^2\,(1 - 0.6)]^{1/2} = 7.16 \text{ db}$$

Then, the required improvement in the signal-to-interference ratio is

$$M_\Omega = 1.64485 \times 7.16 = 11.8 \text{ dB}$$

The required value of Ω_T is then

$$\Omega_T = \mu_\Omega + M_\Omega = 15 \text{ dB} + 11.8 \text{ dB} = 26.8 \text{ dB}$$

6.3 Slow and Fast Fading

When the mobile receiver unit is in a moving automobile, the strength of the received signal varies with time. Decreases in signal strength are known as signal fades and must be allowed for in system design. A typical variation in signal strength is pictured in Figure 6.8. There is a slowly varying component with a typical period, equal to the time taken for the automobile to drive past the width of a building; the responsible phenomenon is known as slow fading and is closely related to shadowing.

Also seen in Figure 6.8 is signal variation with a much shorter time period, on the order of the time taken for the automobile to drive through a wavelength of the radiowave signal. This is known as fast fading and is due to phase interference of signals arriving along different paths (multipath inference).

6.3.1 Slow Fading

With the mobile receiver in an automobile, shadowing varies with time and becomes a wide-sense stationary random process that was discussed in Section 6.1.2. For a mobile communications device in an automobile traveling at a constant speed, v, the autocorrelation function of the shadowing process is a function only of τ, the time taken for the automobile to travel between two different locations along its path, at which the two shadowing random variables are being correlated.

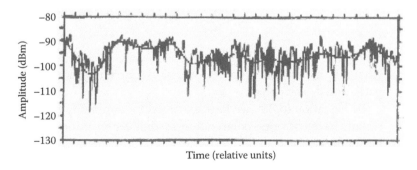

FIGURE 6.8 Signal variation for a mobile receiver in an automobile showing slow and fast fading. Time scale depends on automobile speed.

The autocorrelation function will be exponential in form as given by

$$R_{XX}(\tau) = \exp(-\tau v/r_e) \tag{6.30}$$

where r_e is the autocorrelation distance of the shadowing.

Measurement of r_e allows the system designer to determine how frequently the receiver amplifier will require adjusting to accommodate variation in signal strength caused by slow fading.

6.3.2 Rayleigh Fading

There is an important random process in mobile communications that does not have a log-normal probability distribution function. That process is fast fading due to multipath interference. It is also called fast fading, because the signal amplitude has a Rayleigh probability distribution function as will be described in what follows.

If the mobile receiver in an automobile, it may receive signals along many paths involving reflection from building walls, people, trees, other automobiles, and so forth. Such signals will arrive at the receiver with random phase and when they recombine, fast fluctuations result on a time scale $\sim v/\lambda$, where v is the automobile velocity and λ is the wavelength. There will be other fluctuations due to the variation in shadowing, but these will be on a much slower time scale $\sim v/r_e$ where r_e is the autocorrelation distance of the shadowing, on the scale of the width of a typical building.

When many signal paths of different lengths contribute to the signal electric field, E_s, it becomes a random variable and is given by the phasor

$$E_s = \sum_{n=1}^{N} E_n \exp(j\xi_n) \tag{6.31}$$

where E_n is the field strength of a signal arriving along the nth path, and ξ_n is its phase.

In Equation (6.31), we have assumed for simplicity that all the signals have the same polarization so that vector information can be suppressed.

As shown in Equation (6.31), E_s is a sum of independent identically distributed (i.i.d.) complex random variables. Then, by the central limit theorem as the number of contributing signals becomes large, both the real and imaginary parts of E_s become random variables with Gaussian probability distribution functions and zero means. In other words as N becomes large in Equation (6.31),

$$E_s \rightarrow E_r + jE_i \qquad (6.32)$$

and

$$p_{E_r}(E_r) = \frac{1}{\sqrt{2\pi}\sigma} e^{-E_r^2/2\sigma^2} \qquad (6.33)$$

and

$$p_{Ei}(E_i) = \frac{1}{\sqrt{2\pi}\sigma} e^{-E_i^2/2\sigma^2} \qquad (6.34)$$

The amplitude of E_s is

$$|E_s| = \sqrt{E_i^2 + E_r^2} \equiv E_s \qquad (6.35)$$

which has the probability distribution function

$$p_{\varepsilon s}(E_s) = (E_s/\sigma^2) \exp(-E_s^2/2\sigma^2) \qquad (6.36)$$

Equation (6.36) is a Rayleigh distribution function for the amplitude of the signal electric field.

From Equation (6.36), we can calculate the r.m.s. value of the signal amplitude as

$$E_{r.m.s.} = (EXP[E_s^2])^{1/2} = 1.414\ \sigma \qquad (6.37)$$

and the probability that $E_s < E_1$ is

$$\Pr[E_s < E_1] = \int_0^{E_1} p_{\varepsilon_s}(E)dE = 1 - \exp(-E_1^2/E_{rms}^2) \qquad (6.38)$$

The Rayleigh distribution function given by Equation (6.36) is plotted in Figure 6.9. Note the significant area under the curve in the region $0 < E_s < 1.414\ \sigma$ indicating significant probability of deep fades.

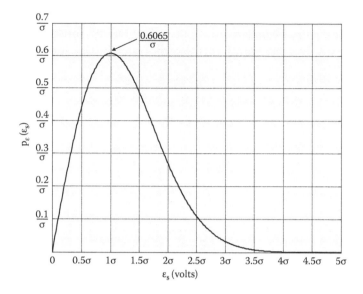

FIGURE 6.9 Rayleigh probability distribution function.

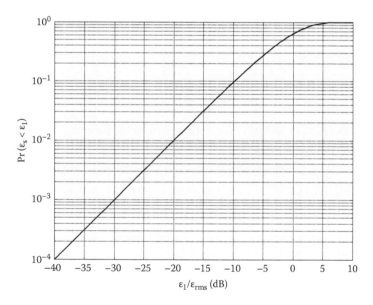

FIGURE 6.10 Probability of Rayleigh fades.

The probability that $E_s < E_1$ (as given by Equation 6.38) is plotted versus $E_1/E_{r.m.s.}$ in dB in Figure 6.10. The way to interpret this curve is that the ordinate gives the probability of a fade equal to or deeper than the corresponding value on the abscissa. For example, the probability of a fade deeper than 40 dB is 10^{-4}.

6.3.3 Margin to Allow for Both Shadowing and Rayleigh Fading

The procedure for calculating sufficient margin to allow for both shadowing and Rayleigh fading is illustrated by the sample problem that follows.

Sample Problem:

Suppose that a mobile phone inside an automobile is subject to Rayleigh fading and is traveling through an area where the shadowing variability is σ_L = 6 dB. If the required probability of call completion is 90%, calculate the margin that must be allowed.

Solution:

To account for shadowing with its log-normal probability density function,

$$Q(M_s/\sigma_L) = p_{out} = 100\% - 90\% = 10\%$$

Then, from Table 6.1, M_s/σ_L = 1.25, and the margin that must be allowed to account for shadowing is M_s = 1.25 × 6 dB = 7.5 dB.

To account for the Rayleigh fading, consult Figure 6.10 that shows that to ensure that deep fades producing outages occur with a probability less than 10%, a margin M_R < 10 dB is required.

Thus, the total required margin is $M = M_s + M_R$ = 17.5 dB.

6.3.4 Bit Error Rates in Digital Communications

We will now consider the effect of stochasticity in path loss on a digital communication channel.

First, consider a channel with log-normal path loss statistics (i.e., with no Rayleigh fading). Such a channel is designated as an all white Gaussian noise (AWGN) channel.

For a digital signal consisting of symbols with duration τ_s, the received energy per symbol is

$$W_s = P_s \tau_s \tag{6.39}$$

If the receiver bandwidth is chosen as $\Delta f = 1/\tau_s$, the noise power received is

$$P_N = N_o \Delta f = N_o/\tau_s \tag{6.40}$$

where for a cell phone system, $N_o = k_B T_o F$.

Then, from Equations (6.39) and (6.40), the signal-to-noise ratio is

$$P_s/P_N \equiv \gamma = W_s/N_o \tag{6.41}$$

For many common types of digital modulation schemes (e.g., binary phase shift keying, BPSK), the probability of an error depends on the signal-to-noise ratio γ as

$$p_e = Q(\sqrt{2\gamma}) = Q\left[(2W_S/N_0)^{1/2}\right] \tag{6.42}$$

Equation (6.42) gives the bit error rate (BER) in an AWGN channel as a function of the signal-to-noise ratio. The Q function has been previously plotted and tabulated in Figure 6.2 and Table 6.1, respectively.

When Rayleigh fading is present, the signal-to-noise ratio becomes a random variable indicated by

$$\gamma = [A_e\tau_s/(2 Z_o N_o)] E_s^2 \tag{6.43}$$

where the signal power into the receiver has been related to E_s by $P_s = [A_e/(2 Z_o)]E_s^2$.

The mean value of the signal-to-noise ratio is

$$\overline{\gamma} \equiv \Gamma = \left[A_e\tau_s/(2Z_0N_0)\right]EXP(E_s^2) = \left[A_e\tau_s/(2Z_0N_0)\right]2\sigma^2$$

$$= A_e\tau_s\sigma^2/(Z_0N_0) \tag{6.44}$$

The probability density function of γ is

$$p_r(\gamma) = p_{\varepsilon_s}(E_s)\left(\frac{dE_s}{d\gamma}\right) \tag{6.45}$$

where $P_{\varepsilon s}$ is given by Equation (6.33).

Now, from Equation (6.43),

$$\frac{d\gamma}{d\varepsilon_s} = [A_e\tau_s/(Z_0N_0)] E_s$$

and inverting gives

$$\frac{d\mathrm{E}_s}{d\gamma} = (\mathrm{Z}_o\mathrm{N}_o)/(\mathrm{A}_e\tau_s\mathrm{E}_s)$$

Then, using Equation (6.36), the probability density function of the signal-to-noise ratio is

$$p_r(r) = [(\mathrm{E}_s/\sigma^2)\exp(\mathrm{E}_s^2/2\sigma^2)]\,[\mathrm{Z}_o\mathrm{N}_o/(\mathrm{A}_e\tau_s\,\mathrm{E}_s)]$$

$$= [(\mathrm{Z}_o\mathrm{N}_o)/(\sigma^2\mathrm{A}_e\tau_s)]\exp(\mathrm{E}_s^2/2\sigma^2) \tag{6.46}$$

Next, using Equations (6.43) and (6.44), Equation (6.46) for the probability density function of γ may be written as

$$p_\gamma(\gamma) = \frac{1}{\Gamma}e^{-\frac{\gamma}{\Gamma}} \tag{6.47}$$

The BER in the Rayleigh channel may then be calculated as

$$BER = EXP[p_e(\gamma)] = \int_0^\infty p_e(\gamma)p_\gamma(\gamma)d\gamma$$

$$= \int_0^\infty Q(\sqrt{2\gamma})\frac{1}{\Gamma}e^{-\gamma/\Gamma}d\gamma = \frac{1}{2}\left[1 - \sqrt{\frac{\Gamma}{1+\Gamma}}\right] \tag{6.48}$$

In Figure 6.11, the BER for a Rayleigh channel as given by Equation (6.48) is plotted together with the BER for a AWGN channel as given by Equation (6.42).

The equivalence of γ and Γ in an AWGN channel has been used. The far smaller value of BER in an AWGN channel should be noted. For example, when the mean signal-to-noise ratio is 10 dB, BER in an AWGN channel is $\sim 4 \times 10^{-5}$ while in the Rayleigh channel it is $\sim 2 \times 10^{-2}$ worse by a factor of 500.

6.3.5 Ricean Fading

There are some communication channels with one dominant signal path plus contributions from many weaker signal paths that

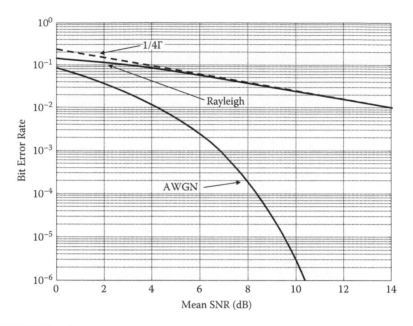

FIGURE 6.11 Comparison of bit error rates in a Rayleigh Channel and in an AWGN Channel.

phase interfere with each other. Such channels are said to follow Ricean statistics. As might be expected, the characteristics of the Ricean channel are intermediate between those of the AWGN channel and the Rayleigh channel.

Consider a total signal given by

$$E_t = E_s e^{j\xi} + E_d e^{j\varsigma} \tag{6.49}$$

where E_s has a Rayleigh p.d.f., E_d is the magnitude of the electric field phasor in the direct (line of sight) signal, and the phase angles ξ and ς are mutually independent and are each uniformly distributed between $-\pi$ and π.

Then, the envelope of the signal $| E_t | = E$ follows the Ricean p.d.f. given by

$$p(E) = (E/\sigma^2) \exp[-(E^2 + E_d^2)/2\sigma^2] I_0(E E_d/\sigma^2) \tag{6.50}$$

where I_0 is the modified Bessel function of the first kind of order zero and $2\sigma^2 = EXP(E_s^2)$.

A Ricean channel is characterized by the ratio of the power in the constant part of the signal to the power in the random part; this ratio is designated by the symbol K:

$$K = E_d^2/(2\,\sigma^2) \tag{6.51}$$

where E_d is the signal voltage in the dominant channel.

$K = 0$ corresponds to the case of a Rayleigh channel, while $K \rightarrow \infty$ corresponds to the AWGN channel.

In Figure 6.12, the Ricean p.d.f. given by Equation (6.50) is plotted for values of $K = 0$, 2, and 8. The transition from a Rayleigh p.d.f. toward a Gaussian p.d.f. with nonzero mean is clearly seen.

6.3.6 Doppler Broadening

A mobile receiver in a moving automobile will experience a Doppler frequency shift in received signals coming from stationary base stations or reflectors. If the wavevector of an incoming signal is **k** and the velocity vector of the automobile is **v**, the Doppler frequency shift is given by

$$\delta f = -\mathbf{k} \cdot \mathbf{v}/(2\,\pi) \tag{6.52}$$

Thus, a signal coming toward the front of the moving automobile (**k** and **v** in opposite directions) will be upshifted in frequency by a maximum amount given by

$$\delta f_m = f_o\,v/c \tag{6.53}$$

where f_o is the unshifted signal frequency.

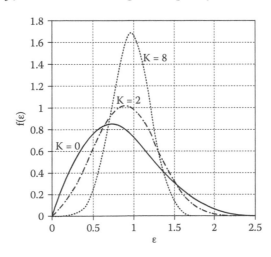

FIGURE 6.12 Ricean probability density function.

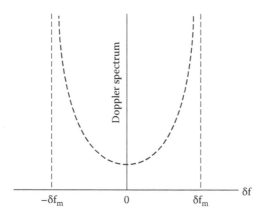

FIGURE 6.13 The classical Doppler spectrum. Abscissa is the displacement from the center frequency.

A signal coming from behind will be downshifted by δf_m. Signals coming toward the automobile at oblique angles will be shifted by smaller amounts than δf_m, in accordance with Equation (6.52).

In a Rayleigh channel, signals are coming at the automobile from all directions, and that results in a continuous frequency spectrum between $f_o - \delta f_m$ and $f_o + \delta f_m$. Assuming that all angles of arrival are equally likely, one can derive the following expression for the power spectral density of the received signal:

$$\Pi(\delta f) = \frac{1.5}{\pi(\delta f_m)\sqrt{1-(\delta f/\delta f_m)^2}} \quad \text{for } |\delta f| < \delta f_m \qquad (6.54)$$

Equation (6.54) is called the "classical Doppler spectrum" and is plotted in Figure 6.13. Doppler frequency shifts outside the band $|\delta f| < \delta f_m$ may occur when the incoming signals originate from a moving source (e.g., reflection from another automobile).

6.4 Wireless Local Area Networks (WLANs)

Wireless local area networks (WLANs) inside buildings, especially those intended to provide Internet access for computers in an office environment, are becoming increasingly common. They allow the system to be based on a single wired connection at the access point or router, while individual computers make wireless contact with that access point. A WLAN is sketched in

Figure 6.14. Each computer has a wireless PC card with a built-in antenna (often a loop antenna). Clearly, there are great advantages in cost and flexibility by exploiting wireless links in this type of network. We will begin consideration of relevant issues for WLANs by describing calculations of path loss inside buildings.

6.4.1 Propagation Losses Inside Buildings

Essentially, the path loss between a transmitting antenna and a receiving antenna inside a building is the free space path loss augmented by excess losses due to the presence of walls and floors through which the signal must propagate. A useful expression for the mean path loss inside a building, $\overline{L}_T$, has been proposed by the International Telecommunications Union (ITU) as

$$\overline{L}_T(dB) = 20\log f_{MHz} + 10n \, \log r + L_{floor}(n_f) - 28 \quad (6.55)$$

where L_{floor} is the floor loss that depends on the number of floors, n_f, through which the signal travels, and r is the range in meters.

In Equation (6.55), wall losses are not explicitly included. The effects of walls and other obstructions are accounted for by allowing the range exponent, n, to be larger than 2.

For an office building environment, the ITU model reports values of n in the range $3.3 > n > 2.8$ for values of signal frequency between 900 MHz and 4 GHz. Taking a ballpark value of $n = 3$ would seem to be reasonable in this frequency range in office buildings. ITU data also suggest that for residential buildings, $n = 2.8$

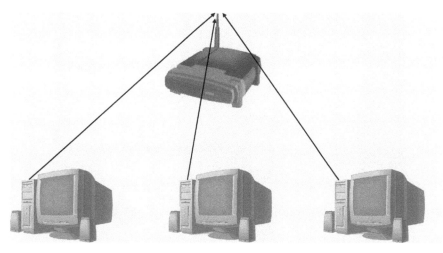

FIGURE 6.14 Wireless local area network (WLAN) of computers.

is appropriate. For department store environments characterized by large open spaces with few walls, the data suggest that an appropriate value of the range exponent is $n = 2.2$.

The floor loss in office buildings for the type of frequency range discussed in the preceding paragraph is 9 dB for one floor. For $2 \leqslant n_f \lesssim 7$, the following expression may be used to estimate floor loss:

$$L_{\text{floor}} \, (\text{dB}) = 15 + 4 \, (n_f - 1) \tag{6.56}$$

Equation (6.56) gives an overestimate when the number of floors is eight or more, because alternate, lower loss, wave paths exist. For example, a path might involve passing out a window, reflecting off the wall of a neighboring building, and then coming back in through a window on a different floor. Such a path may provide lower loss than passing through all the floors when eight floors or more separate the transmitter from the receiver.

This alternate wave path is pictured in Figure 6.15. The total path loss associated with each path pictured in Figure 6.15, versus the number of floors, is plotted in Figure 6.16. Bear in mind that as the number of floors increases, the range also increases. Thus, both the second term and the third term on the right-hand side of Equation (6.55) become larger.

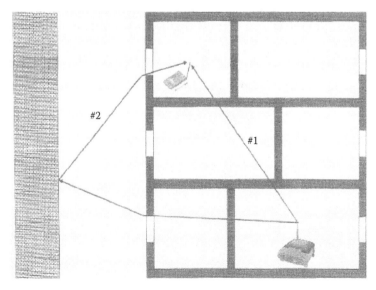

FIGURE 6.15 Alternative paths for propagation between floors.

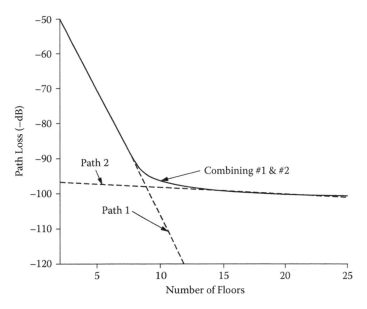

FIGURE 6.16 Variation of path loss with number of floors. Floor height is 4 m; building width is 30 m; distance to adjacent building is 30 m; frequency is 900 MHz.

For residential buildings and department stores, the floor loss is not as large as indicated by Equation (6.55). For residential buildings, the ITU gives

$$L_{floor} \ (dB) = 4 \ n_f \qquad (6.57)$$

For department stores, the ITU expression is

$$L_{floor} \ (dB) = 6 + 3 \ (n_f - 1) \qquad (6.58)$$

Finally, in this section on path loss inside buildings, one must take into account the fact that in any given building, there will be variation in the spacing between walls, the thickness of the walls, the material from which the walls are made, and the arrangement of office furniture. These factors are random. Thus, path loss inside a building is subject to shadowing in the same way as are losses in an outside environment. As before, shadowing is a random variable with a log-normal, zero-mean p.d.f. An important parameter is σ_s, the standard deviation of the p.d.f.

As discussed in Section 6.1, one must choose the maximum probability of error, p_e, that can be tolerated. Then, by setting

$$p_e = Q \ (M/\sigma_s) \qquad (6.59)$$

one can determine the required margin, M. The path loss to be allowed for, in the design of the indoor WLAN, is then

$$L(dB) = \bar{L}_T(dB) + M \, (dB) \tag{6.60}$$

6.4.2 Standards for WLANs

The Institute of Electrical and Electronic Engineers (IEEE) has attempted to standardize WLAN equipment. The standards will undoubtedly change over the years. But the IEEE standard designations as of 2006 are summarized in Table 6.2.

WiFi equipment (802.11b) is the most popular, because a large variety of inexpensive components are available that correspond to this standard. However, the operating frequency of 2.4 GHz is shared with other equipment, such as microwave ovens and cordless telephones, and interference from these appliances may be encountered.

The 802.11a equipment operates at 5 GHz where interference problems may be expected to be less severe. The link speed is faster than in WiFi systems. Also, there are eight channels instead of three.

The most flexible system is built to the 802.11g standard. It has dual operating frequency bands (centered at 2.4 and 5 GHz), and so is compatible with both 802.11b and 802.11a equipment. It also has the faster link speed capability as in 802.11a.

Note that the range of the systems operating at 5 GHz is smaller than for systems operating at 2.4 GHz, as would be expected. Also, at a given operating frequency, outdoor range (where path loss $L \sim 20 \log r$) is much larger than indoor range (where $L \sim 30 \log r$).

TABLE 6.2   Institute of Electrical and Electronic Engineers (IEEE) Standards for Wireless Local Area Networks (WLANs)

Standards Designation	Range	Center Frequency, f_o	Number of Channels	Maximum Link Speed
802.11b, WiFi	48 m indoors 600 m outdoors	2.4 GHz	3	11 Mbps
802.11a	18 m indoors 480 m outdoors	5 GHz	8	54 Mbps
802.11g	Similar to 802.11a or 802.11b depending on frequency	1.4 GHz and 5 GHz	3	54 Mbps

6.4.3 Sharing WLAN Resources

Finally, we note that many schemes are used for sharing WLAN resources. In companies that occupy more than a single building, a wireless bridge can be established between buildings so that files may be shared. Such a wireless bridge between two buildings is pictured in Figure 6.17, where high-gain parabolic reflector antennas are used to maintain data security.

If two different companies occupying the same building wish to set up separate WLAN networks operating at the same frequency with minimal interference, this may be done if the two networks are separated by many floors as pictured in Figure 6.18. The system design would rely on floor loss so that in-network signals would be larger than interfering signals from the second network by at least 20 dB (i.e., $C/I > 20$ dB). As discussed in

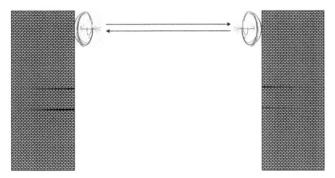

FIGURE 6.17   Wireless bridge.

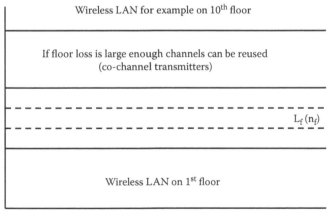

Wireless LAN for example on 10th floor

If floor loss is large enough channels can be reused
(co-channel transmitters)

$L_f(n_f)$

Wireless LAN on 1st floor

In channel signal to interference ratio $C/I \geq 20$ dB

FIGURE 6.18   Co-channel reuse and interference.

Chapter 1, similar criterion were used in cell phone systems where frequency bands were reused in different cells.

Problems

6.1. A cell phone system in an urban area is to be designed to allow reception at the cell edge with a 96% success rate. The base station antenna height is 50 m, the lowest mobile antenna height is 1.5 m, the frequency is 880 MHz, the standard deviation of the shadowing is 12 dB, and the maximum acceptable path loss is 140 dB. Calculate the maximum cell radius.

6.2. Repeat the calculation in Problem 6.1 with the added provision that the signal must be accessible inside buildings. The building loss has an average value of 11 dB and a standard deviation of 6 dB.

6.3. A cell phone system is to provide a 90% call success rate at the cell edge with a location variability of 8 dB. What margin is required? What is the probability of a successful call averaged over an entire cell? Assume the path loss exponent is four.

6.4. Given that the autocorrelation function of shadowing falls to 1/e of its initial value at a distance of 60 m, estimate the time intervals at which signal strength needs to be adjusted for a cell phone in an automobile traveling at 35 miles per hour.

6.5. If a mobile phone in an automobile traveling at 50 m/s approaches a second automobile coming toward it at a speed of 40 m/s, what is the maximum Doppler shift the mobile phone can experience if it receives signals from a base station transmitting at 900 MHz, indirectly by way of reflection from the second automobile?

6.6. An automobile is driven through an urban area where the variability of the shadowing is 10 dB. Calculate the margin that must be allowed to ensure that the probability of successfully completing a call is 95%, accounting for both slow fading (shadowing) and fast fading (Rayleigh fading).

6.7. In calculating the Rayleigh error performance of a BPSK system, it was assumed that fading was constant over one bit duration. Up to what vehicle speed is this assumption valid given that the bit rate is 250 bps and the carrier frequency is 880 MHz?

6.8. A mobile receiver using BPSK in an automobile suffers a BER of 10^{-5} when the automobile is stationary. To what new value will the BER rise due to Rayleigh fading when the automobile is moving, if there is no error-correcting scheme in use?

6.9. In an urban area, satisfactory cell phone service requires a signal-to-interference ratio, Ω, of at least 15 dB with a probability of 95%. The standard deviation of the shadowing is σ_L = 8 dB.

Calculate the threshold value of the signal-to-inference ratio, Ω_T, for which a cell phone system must be designed for two cases as follows:

a. The cross-correlation of shadowing along the in-cell path and the interfering is $R_{12} = 1$.

b. The cross-correlation of shadowing along the in-cell path and the interfering is $R_{12} = 0$.

6.10. Calculate the maximum number of floors that can separate a computer and an access point in an office building wireless LAN operating at 2.4 GHz, with a maximum horizontal separation of 20 m and a height of 4 m per story if the maximum acceptable path loss is 110 dB. Assume that shadowing effects are negligible.

6.11. Repeat Problem 6.10, accounting for shadowing with a variability of 4 dB and requiring a 99% connection success rate.

Bibliography

1. J.G. Proakis, *Digital Communications*. 2nd ed. (McGraw-Hill, New York, 1989).

2. S.O. Rice, "Mathematical Analysis of a Sine Wave Plus Random Noise," *Bell System Technical Journal* 27 (1948): 109–157.

3. K. Siwiak, *Radiowave Propagation and Antennas for Personal Communications* (Artech House, Boston, Massachusetts, 1998).

4. T.L. Kelley, *The Kelley Statistical Tables* (Harvard University Press, Cambridge, Massachusetts, 1948).

5. V. Graziano, "Propagation Correlations at 900 MHz," *IEEE Transactions on Vehicular Technology* 27 (1978): 182–189.

6. S.R. Saunders, *Antennas and Propagation for Wireless Communication Systems* (John Wiley & Sons, Chichester, UK, 1999), 180–222.

7. G.R. Grimmett and D.R. Stirzaker, *Probability and Random Processes*. 2nd ed. (Clarendon Press, Oxford, UK, 1992).

CHAPTER 7

Tropospheric and Ionospheric Effects in Long-Range Communications

In the preceding chapters, we concentrated on cell phone system communications where ranges of a few kilometers are of interest, or on wireless local area networks (WLANs) where ranges are a few tens of meters. However, from its onset, wireless communications has demonstrated its capacity for long-range communications across vast oceans and, nowadays, across interplanetary space. In this chapter, we explore how the physical properties of the inner layer of the earth's atmosphere (the troposphere) and the outer layer (the ionosphere) may be used to extend the range of communications beyond the line-of-sight limit of communications before the discovery of radio frequency (RF) waves. In Chapter 8, we will describe the role of artificial satellites of the earth in extending the frequency range of worldwide wireless communications.

7.1 Extending the Range Using Tropospheric Refraction

7.1.1 Limit on Line-of-Sight Communications

Before the advent of RF waves, long-range communications, using flashing lights or using signal flag and telescopes, were limited by the bulge of the spherical earth. This limit can be evaluated by considering Figure 7.1, where a transmitting antenna at a height h_A above the surface of the earth transmits a signal to the horizon. At the horizon, the direct path from the transmitter grazes the surface of the earth and is normal to the earth's radius.

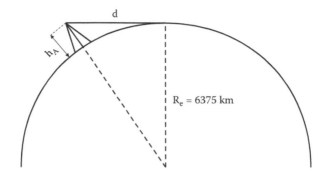

FIGURE 7.1 Geometry for calculating the distance from an antenna to a line-of-sight (LOS) horizon.

For the right-angle triangle shown in Figure 7.1,

$$(R_e + h_A)^2 = R_e^2 + d^2$$

or

$$R_e^2 + 2R_e h_A + h_A^2 = R_e^2 + d^2 \tag{7.1}$$

In Equation (7.1), the antenna height, h_A, is much less than the earth's radius, R_e = 6375 km. Thus, we can neglect h_A^2 compared with R_e^2 in Equation (7.1) and obtain an approximate expression for the range to the horizon as

$$d \approx \sqrt{2R_e h_A} \tag{7.2}$$

For example, if antenna height were 100 m, Equation (7.2) gives the maximum line-of-sight range as

$$d = \sqrt{2 \times 6.375 \times 10^6 \times 100} = 36 km$$

As will be seen below, this range can be extended somewhat if tropospheric refraction is exploited. In the troposphere, the density of the air decreases with increasing altitude. The resulting decease in refractive index, $n = (\varepsilon_r \, \mu_r)^{1/2}$, with increasing altitude, h, causes the path of a radio wave to be bent toward the ground. Then, as shown in Figure 7.2, when a wave is launched in the locally horizontal direction, it may propagate to a point on the earth beyond the horizon.

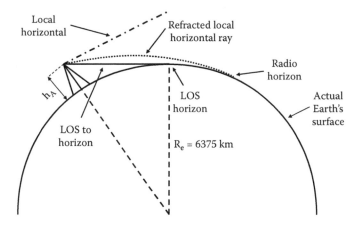

FIGURE 7.2 Comparison of line-of-sight (LOS) horizon with radio horizon.

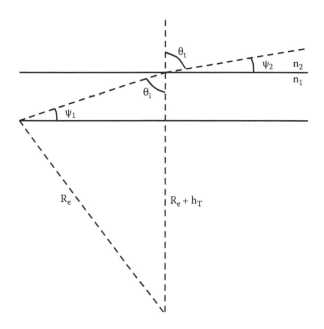

FIGURE 7.3 Geometry for deriving Bouguer's law.

7.1.2 Bouguer's Law for Refraction by Tropospheric Layers

Consider what happens when there is a sharp decrease in refractive index at an altitude $h = h_T$ as depicted in Figure 7.3, where n_1 is the refractive index for $h < h_T$, and n_2 is the refractive index for $h > h_T$. Then, a wave launched with elevation angle ψ_1 that is incident on the interface between the two dielectric layers with angle of incidence, θ_i, will be refracted so that the angle of transmission,

θ_t, will be larger than θ_i. Applying the law of sines to the triangle in Figure 7.3 gives

$$\frac{R_e + h_T}{\sin(\Psi_1 + \frac{\pi}{2})} = \frac{R_e}{\sin\theta_i}$$

or

$$\sin\theta_i = \frac{R_e}{R_e + h_T}\sin(\Psi_1 + \frac{\pi}{2}) = \frac{R_e}{R_e + h_T}\cos(\Psi_1) \qquad (7.3)$$

Also, from Snell's law of refraction,

$$\sin\theta_i = \frac{n_2}{n_1}\sin\theta_t \qquad (7.4)$$

Then, equating the right-hand side of Equation (7.3) to the right-hand side of Equation (7.4) gives

$$\frac{R_e}{R_e + h_T}\cos(\Psi_1) = \frac{n_2}{n_1}\sin\theta_t = \frac{n_2}{n_1}\cos(\Psi_2)$$

or

$$\frac{\cos(\Psi_1)}{\cos(\Psi_2)} = \frac{R_e + h_T}{R_e}\frac{n_2}{n_1} \qquad (7.5)$$

where ψ_1 is the elevation angle of the ray path in medium 1, and ψ_2 is the elevation angle of the refracted ray path in medium 2.

Equation (7.5) is known as Bouguer's law. In order for the ray path to bend back toward the earth, one requires $\psi_2 < \psi_1$, or from Equation (7.5),

$$\frac{R + h_T}{R}\frac{n_2}{n_1} < 1 \qquad (7.6)$$

7.1.3 Increase in Range Due to Tropospheric Refraction

In the troposphere, the refractive index decreases continuously with altitude. One can appreciate that this might result in a curving of the ray path toward the earth by considering Figure 7.4, where the refractive effect of many thin layers is pictured. Each layer has an incrementally smaller value of refractive index than the layer immediately below it. In the limit of a continuously decreasing refractive index, the ray path curves toward the earth with radius of curvature equal to $(-dn/dh)^{-1}$. As was shown in Figure 7.2, the curvature of the ray path may result in an extended distance for communications beyond the horizon. To good approximation the extended distance is given by

$$d = \sqrt{2k_e R_e h_A} \qquad (7.7)$$

The enhancement factor, k_e, is called the "effective earth radius factor" and may be calculated from the expression

$$k_e = \frac{1}{1 + R_e \dfrac{dn}{dh}} \qquad (7.8)$$

Because dn/dh is negative, it can be seen from Equation (7.8) that $k_e > 1$.

The refractive index of the earth's troposphere is given under average conditions by

$$n\,(\text{troposphere}) = 1 + N_t \times 10^{-6} \qquad (7.9)$$

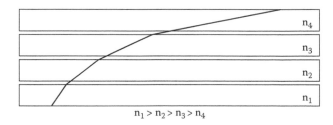

$$n_1 > n_2 > n_3 > n_4$$

FIGURE 7.4 Bending of a ray path by successive dielectric layers with decreasing refractive index.

where the "*N*-number," $N_t = 315 \exp(-h/H)$ and the e-folding distance is $H = 7.35$ km.

If h « *H*, the *N*-number

$$N_t \approx 315 \, (1 - h/H)$$

and

$$\frac{dn}{dh} = -\frac{315 \times 10^{-6}}{H} = -43 \times 10^{-6} \, km^{-1}$$

If we now insert this value into Equation (7.8), we get for the effective earth radius factor

$$k_e = \frac{1}{1 - 6375 \times (43 \times 10^{-6})} = 1.38$$

The enhanced distance for beyond the horizon communications may next be calculated from Equation (7.7) as

$$d = \sqrt{2k_e R_e h_A} = 42 \text{ km}$$

where we have again assumed that $h_A = 100$ m.

This calculated value of 42 km should be compared with the previously calculated value of the distance to the horizon, which was 36 km.

In 1900, Marconi was able to communicate across the English Channel over distances exceeding the "earth bulge limit" due to tropospheric refraction. He was unaware of the nature of the physical phenomenon that accounted for his success. Nevertheless, he was encouraged to attempt in the following year wireless communication across the Atlantic Ocean from Wales to Newfoundland. Despite the prevalent skepticism among the scientific community of that time, his trans-Atlantic experiment also succeeded.

Clearly, from our preceding calculation, tropospheric refraction could not account for a range of several thousand kilometers. A different physical phenomenon was operational. Within a year of Marconi's demonstration of trans-Atlantic communications, the existence of an electrically charged layer of particles at an elevated altitude was postulated to be the reflector of the trans-Atlantic radio waves. This charged layer is now called the ionosphere. It will be described in the next section.

7.2 Long-Range Communications by Ionospheric Reflection

7.2.1 The Ionospheric Plasma

At an altitude of roughly 300 km above the earth's surface, the density of atmospheric gases is such that the ultraviolet rays in sunlight cause ionization and the formation of a plasma layer. A plasma layer is a macroscopically neutral layer of ionized gases in which the charge density of the positive ions is balanced by the charger density of electrons.

To evaluate how an RF wave interacts with plasma, it is appropriate to assume that the relatively massive ions are immobile, while the much lighter electrons move in response to the fields of the radiowave. The equation of motion for an electron is

$$\frac{d[m\vec{v}(t)]}{dt} = -e[\vec{E}(t) + \vec{v}(t) \times \mu_o \vec{H}(t)] \tag{7.10}$$

where m is electron mass, e is the magnitude of electron charge, and $v(t)$ is electron velocity.

Assuming that $v \ll c$ (nonrelativistic limit), the electron mass closely appoaches its rest mass value equal to 9.11×10^{-31} kg.

Also, we note that

$$\left| \vec{v} \times \mu_0 \vec{H} \right| \leq v\mu_0 H = v\mu_0 \frac{E}{Z_0} = \frac{v}{c} E \ll E$$

Thus, we can neglect the second term in the brackets on the right-hand side of Equation (7.10). This equation then reduces to

$$m\frac{d\vec{v}(t)}{dt} = -e\vec{E}(t) \tag{7.11}$$

In phasor notation, Equation (7.11) is written as

$$j\omega\vec{v} = -e\vec{E}$$

or

$$\vec{v} = \frac{-e\vec{E}}{j\omega m} \tag{7.12}$$

Now using the Maxwell–Ampere law and realizing that current density is due to the movement of the free electrons (convective current density), we get

$$\nabla \times \vec{H} = j\omega\varepsilon_0\vec{E} + \vec{J} = j\omega\varepsilon_0\vec{E} + n_e e\vec{v} \tag{7.13}$$

where n_e is the electron number density.

Next, substituting from Equation (7.12) into Equation (7.13) gives

$$\nabla \times \vec{H} = j\omega\varepsilon_0\vec{E} + \frac{n_e e^2}{j\omega m}\vec{E} = j\omega\varepsilon_0\vec{E}\left(1 - \frac{n_e e^2}{\varepsilon_0 m\omega^2}\right) \tag{7.14}$$

Thus, we see from Equation (7.14) that the plasma may be represented by a relative permittivity:

$$\varepsilon_r = \left(1 - \frac{n_e e^2}{\varepsilon_0 m\omega^2}\right) = 1 - \frac{\omega_p^2}{\omega^2} \tag{7.15}$$

where we have defined the angular plasma frequency as

$$\omega_p = \sqrt{\frac{n_e e^2}{m\varepsilon_0}} \tag{7.16}$$

All the terms under the square root sign in Equation (7.16) are constants. Thus, this equation may be rewritten in the convenient form

$$f_p = 8.98\sqrt{n_e} \tag{7.17}$$

where the plasma frequency, f_p, is in Hz and the electron number density, n_e, is in m^{-3}.

The plasma frequency is a natural resonant oscillation frequency of the layer of ions and electrons.

7.2.2 Radio Frequency (RF) Wave Interaction with Plasma

For a plane wave propagating through a plasma, the relation between the wavenumber, k, and the angular frequency, ω, is simply given by

$$k^2 = \omega^2 \mu \varepsilon = \omega^2 \mu_0 \varepsilon_0 \varepsilon_r = \omega^2 \mu_0 \varepsilon_0 \left(1 - \frac{\omega_p^2}{\omega^2}\right) \qquad (7.18)$$

Equation (7.18) is plotted in Figure 7.5. In this figure, it may be seen that no values of k are plotted for frequencies lower than the plasma frequency because, as may be seen from Equation (7.18), k would be imaginary in this frequency range. Thus, the plasma acts as a high-pass filter and only allows waves with a frequency larger than the plasma frequency to propagate through it.

To understand what happens when waves in air are incident on the plasma layer as pictured in Figure 7.6, it is necessary to recall that the relative permittivity of the plasma is less than one. Thus, we have the case of a wave in a region of relatively high permittivity (medium #1, air) incident on a region of lower permittivity (medium #2, plasma). In such a case, there will be complete specular reflection of the wave when the angle of incidence, θ_I, exceeds the critical angle, θ_c, which is defined by

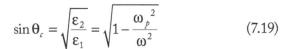

$$\sin \theta_c = \sqrt{\frac{\varepsilon_2}{\varepsilon_1}} = \sqrt{1 - \frac{\omega_p^2}{\omega^2}} \qquad (7.19)$$

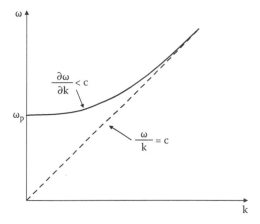

FIGURE 7.5 Dispersion curve for wave propagation in a plasma.

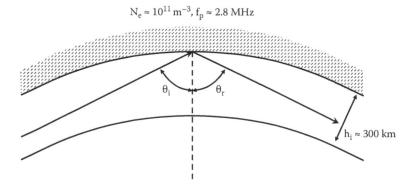

FIGURE 7.6 RF wave incident on the ionospheric plasma.

Another way of stating the concept, embodied by Equation (7.19), is that the wave will be completely reflected at the interface if

$$\sin\theta_i \geq \sqrt{1 - \frac{\omega_p^{\,2}}{\omega^2}}$$

Equivalently, by rearranging the terms in the inequality, one can expect complete reflection if

$$\omega \leq \omega_p/\cos\theta_i \qquad (7.20)$$

A simple picture of ionospheric reflection is that the radio wave propagates upward through a region of increasing plasma density until

$$\omega_p = \omega\cos\theta_i \qquad (7.21)$$

At that altitude, complete reflection takes place.

The maximum usable frequency in a communications system depending on ionospheric reflection will then depend on the maximum value of plasma frequency in the ionosphere and the maximum value of the angle of incidence. From Equation (7.21), the maximum usable frequency is given by

$$f_{muf} = f_{p,max} / \cos(\theta_{i,max}) \qquad (7.22)$$

The maximum value of plasma frequency can be determined from plots of electron number density versus altitude, such as those shown in Figure 7.7. The attainable plasma frequency maximum clearly depends on factors such as the time of day or night. The nighttime maximum is smaller than the daytime maximum, but it occurs at higher altitude. This implies that nighttime communications will be limited to lower frequency but will have a longer maximum range.

To calculate the maximum value of the angle of incidence, assume a ray path normal to the earth's radius at the location of the transmitting antenna as shown in Figure 7.8. From the right angle triangle in this figure we see that

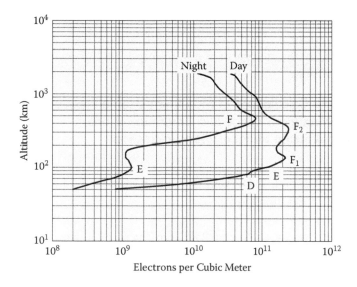

FIGURE 7.7 Maps of ionospheric electron density versus altitude.

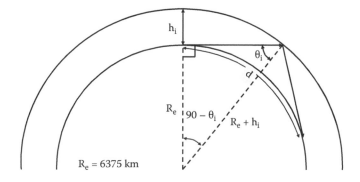

FIGURE 7.8 Geometry for determining maximum range in communication with single ionospheric reflection.

$$\sin \theta_{i,\max} = \frac{R_e}{R_e + h_i} \qquad (7.23)$$

where h_i is the height of the ionosphere at which wave reflection occurs.

Also, from Figure 7.8, the maximum range for communication by ionospheric reflection is seen to be

$$d_{\max} = 2R_e \left[\frac{90^0 - \theta_i}{57.3^0} \right] \qquad (7.24)$$

Equations (7.22), (7.23), and (7.24) are the basic equations to be used in the design of a communication link using ionospheric reflection.

7.2.3 Sample Calculations of Maximum Usable Frequency and Maximum Range in a Communications System Based on Ionospheric Reflection

Example 1: Daytime Reflection from the F1 Layer
From Figure 7.7,

$$N_{e,\max} = 2 \times 10^{11} \mathrm{m}^{-3} \text{ at } h_i = 120 \text{ km}$$

This value of maximum electron number density corresponds to a maximum value of plasma frequency given by

$$f_{p,max} = 8.98\sqrt{2 \times 10^{11}} = 4.02 \times 10^6 \, Hz$$

The maximum angle of incidence is determined by the height of the reflecting layer (viz.,) h_i = 120 km. From Equation (7.23),

$$\theta_{i,\max} = \sin^{-1} \frac{6375}{6375 + 120} = 79^0$$

Then, from Equation (7.22) the maximum usable frequency is

$$f_{muf} = \frac{4.02 \times 10^6}{\cos 79^0} = 21\text{MHz}$$

And from Equation (7.24), the maximum range is

$$d_{MAX} = 2 \times 6375\text{km} \times \frac{90^0 - 79^0}{57.3^0} = 2248\text{km}$$

Example 2: Nighttime Reflection from the F Layer
From Figure 7.7,

$$N_{e,\max} = 8 \times 10^{10}\, m^{-3} \text{ at } h_i = 450\text{km}$$

The corresponding value of maximum plasma frequency is

$$f_{p,\max} = 8.98\sqrt{8 \times 10^{10}} = 2.54 \times 10^6\,\text{Hz}$$

The maximum angle of incidence corresponding to $h_i = 120$ km is given by Equation (7.23) as

$$\theta_{i,\max} = \sin^{-1} \frac{6375}{6375 + 450} = 69.1^0$$

Then the maximum usable frequency can be determined using Equation (7.22) as

$$f_{muf} = \frac{2.54 \times 10^6}{\cos 69.1^0} = 7.1 MHz$$

The maximum range can be determined by using Equation (7.24) and is

$$d_{\max} = 2 \times 6375\text{km} \times \frac{90^0 - 69.1^0}{57.3^0} = 4650\text{km}$$

Thus, as anticipated, the nighttime maximum usable frequency is smaller (7.1 MHz versus 21 MHz) while the maximum range is significantly larger (4650 km versus 2248 km). This is the reason that some distant radio stations can be heard only at night.

7.3 Propagation through the Ionosphere

When the signal frequency is sufficiently large compared with the plasma frequency the signal wave will pass through the ionosphere. This occurs when

$$f > f_{p,max} / \cos(\theta_{i,max})$$

This higher frequency range is important, for example, in satellite communications (SATCOM). The signal wave, although not reflected, will still be influenced by the presence of the plasma; specifically, the signal will experience time delay, dispersion, and change in polarization direction.

7.3.1 Time Delay of a Wave Passing through the Ionosphere

A signal carried by an electromagnetic wave travels at the group velocity of the wave. Due to the presence of the ionosphere, the group velocity of a wave is slower than in free space. The group velocity is given in general by

$$v_g = \partial \omega / \partial k \tag{7.25}$$

Specifically, for wave propagation through a plasma, one may calculate from the plasma dispersion relation, given by Equation (7.18), that

$$v_g = c\sqrt{1 - (f_p^2 / f^2)} \tag{7.26}$$

Thus, the incremental time delay introduced by the plasma when the wave propagates through an incremental length, dz, is

$$dt_d = (dz/v_g) - (dz/c) \tag{7.27}$$

Then, substituting Equation (7.26) into Equation (7.27) gives

$$dt_d = \frac{dz}{c} \left[\left(1 - \frac{f_p^2}{f^2} \right)^{-1/2} - 1 \right] \tag{7.28}$$

For plasma frequency much smaller than the signal frequency, Equation (7.28) to good approximation becomes

$$dt_d = \frac{1}{2}\frac{dz}{c}\frac{f_p^2}{f^2} \tag{7.29}$$

The plasma frequency may be related to the electron density by Equation (7.17), so that in mks units, Equation (7.29) may be written

$$dt_d = \frac{40.3}{cf^2}n_e(z)dz \tag{7.30}$$

To obtain the total time delay, one integrates Equation (7.30) along the entire path to get

$$t_d = \frac{40.3}{cf^2}n_T \tag{7.31}$$

where f is in Hz and the total electron content is defined by

$$n_T = \int_{wavepath} n_e(z)dz \tag{7.32}$$

which has dimensions of (meters)$^{-2}$.

The total electron content is an important parameter in describing a number of ionospheric effects. For a zenith path, through the entire ionosphere, n_T is in the range 10^{16} to 10^{18} m^{-2}.

7.3.2 Dispersion of a Wave Passing through the Ionosphere

The time delay described by Equation (7.31) is a function of frequency. Thus, different frequency components of the signal will experience different delays, and distortion of the signal will result. This phenomenon, which is especially disturbing for wideband signals, is called *dispersion*.

A measure of dispersion is dt_d/df, which may be calculated from Equation (7.31) as

$$\frac{dt_d}{df} = -\frac{80.6}{cf^3}n_T \tag{7.33}$$

where mks units are used.

When a signal occupies a bandwidth Δf, the spread in time of arrival of the spectral components of the signal will be

$$\Delta t_d = -\frac{80.6}{cf^3}n_T\Delta f \tag{7.34}$$

The highest frequency component will arrive at the receiver first.

7.3.3 Faraday Rotation of the Direction of Polarization in the Ionosphere

A linearly polarized wave propagating through the ionosphere will have the direction of its electric field rotated when it propagates through the ionosphere, due to interaction with free electrons in the earth's magnetic field. As indicated in Figure 7.9, a wave propagating in a magnetized plasma splits into two modes, each with distinct propagation characteristics: the ordinary mode with electric field parallel to the constant magnetic field, and the extraordinary mode with its electric field normal to the constant magnetic field. The ordinary mode is relatively unaffected by the presence of the plasma. It is the extraordinary mode that will have its direction of polarization rotated.

To analyze this phenomenon, we will make the assumption that the wavevector, **k**, is parallel to the constant magnetic flux density vector, $\mathbf{B}_e = B_e\,\mathbf{a}_z$. The permittivity of the collisionless, magnetized plasma is then a tensor—that is,

$$\mathbf{D} = \varepsilon_{xy}\cdot\mathbf{E} \tag{7.35}$$

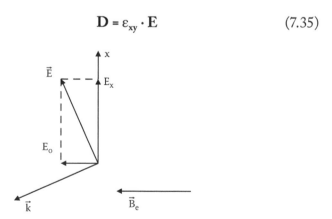

FIGURE 7.9 Wave propagating in a magnetized plasma split into an ordinary mode (E_o) and an extraordinary mode (E_x).

where

$$\varepsilon_{xy} = \varepsilon_0 \begin{bmatrix} \kappa_1 & -j\kappa_2 & 0 \\ j\kappa_2 & \kappa_1 & 0 \\ 0 & 0 & \kappa_1 \end{bmatrix} \qquad (7.36)$$

In this tensor,

$$\kappa_1 = 1 - (\omega_p/\omega)^2 \qquad (7.37)$$

and

$$\kappa_2 = \omega_p^2 \, \omega_c / [\omega \, (\omega^2 - \omega_c^2)] \qquad (7.38)$$

The new resonant frequency, ω_c, is the electron cyclotron frequency in the presence of a constant magnetic flux density, B_e, and is defined by

$$\omega_c = e \, B_e/m \qquad (7.39)$$

An electron in the presence of a constant magnetic field will rotate in a plane normal to that field at the electron cyclotron frequency as shown in Figure 7.10.

For a value of the earth's magnetic flux density, $B_e = 7 \times 10^{-5}$ Tesla, the electron cyclotron frequency is 2 megahertz. In Equations (7.37) and (7.38), it was assumed that the electron cyclotron frequency was much less than the plasma frequency.

Equation (7.36) may be expanded as

$$D_x = \varepsilon_0 \, \kappa_1 \, E_x - j \, \varepsilon_0 \, \kappa_2 \, E_y \qquad (7.40a)$$

$$D_y = \varepsilon_0 \, \kappa_1 \, E_y + j \, \varepsilon_0 \, \kappa_2 \, E_x \qquad (7.40b)$$

$$D_z = \varepsilon_0 \, \kappa_1 \, E_z \qquad (7.40c)$$

Now, consider a plane wave with extraordinary polarization propagating through the magnetized plasma with a wavevector parallel to the constant magnetic field (i.e., $\mathbf{k} = k \, \mathbf{a}_z$). The geometry is pictured in Figure 7.11. Maxwell's equations describing this wave are

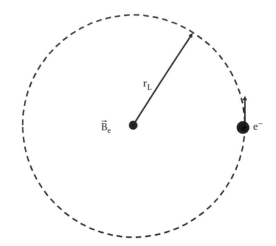

FIGURE 7.10 Electron orbit in a plane normal to a constant magnetic field with a radius equal to the Larmor radius, r_L.

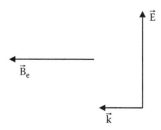

FIGURE 7.11 Wave with "extraordinary" polarization propagating through a magnetized plasma.

$$\nabla \times \boldsymbol{E} = -j\,\omega\,\mu_o\,\boldsymbol{H} \qquad (7.41a)$$

$$\nabla \times \boldsymbol{H} = j\,\omega\,\varepsilon_{xy} \cdot \boldsymbol{E} \qquad (7.41b)$$

Assume that solutions of Equation (7.38) are of the form

$$\boldsymbol{E} = \boldsymbol{E}_o\,e^{-j\beta z} \qquad (7.42a)$$

and

$$\boldsymbol{H} = \boldsymbol{H}_o\,e^{-j\beta z} \qquad (7.42b)$$

where $\boldsymbol{E}_o$ and $\boldsymbol{H}_o$ are constant vectors. Then, Equations (7.41a) and (7.42b) can be combined to get a wave equation.

Taking the curl of Equation (7.41a) and substituting from Equation (7.42b) gives

$$\nabla \times \nabla \times \boldsymbol{E} = -j\, \omega\, \mu_o \nabla \times \boldsymbol{H} = \omega^2\, \mu_o\, \varepsilon_{xy} \cdot \boldsymbol{E}$$

or

$$\nabla \nabla. \ \boldsymbol{E} - \nabla^2\, \boldsymbol{E} = \omega^2\, \mu_o\, \varepsilon_{xy} \cdot \boldsymbol{E} \qquad (7.43)$$

Now, because the effect of the plasma electrons has been absorbed into the dielectric constant tensor, effectively, there is no free charge and Ñ. $\boldsymbol{E}$ = 0. Then Equation (7.43) may be written as

$$\beta^2\, \boldsymbol{E}_o = \omega^2\, \mu_o\, \varepsilon_{xy} \cdot \boldsymbol{E}_o \qquad (7.44)$$

Using Equation (7.36), Equation (7.44) may be expanded as

$$\beta^2\, E_{ox} = \omega^2\, \mu_o\, \varepsilon_o\, \kappa_1\, E_{ox} - j\, \omega^2\, \mu_o\, \varepsilon_o\, \kappa_2\, E_{oy} \qquad (7.45a)$$

and

$$\beta^2\, E_{oy} = j\, \omega^2\, \mu_o\, \varepsilon_o\, \kappa_2\, E_{ox} + \omega^2\, \mu_o\, \varepsilon_o\, \kappa_1\, E_{oy} \qquad (7.45b)$$

In matrix form, Equations (7.45) may be written as

$$\begin{bmatrix} \beta^2 - k_o^2 \kappa_1 & jk_o^2 \kappa_2 \\ -jk_o^2 \kappa_2 & \beta^2 - k_o^2 \kappa_1 \end{bmatrix} \begin{bmatrix} E_{ox} \\ E_{oy} \end{bmatrix} = 0 \qquad (7.46)$$

Equation (7.46) is a homogeneous equation in E_{ox} and E_{oy}. For nontrivial solutions, the determinant of the coefficient matrix must equal zero; that is,

$$(\beta^2 - k_o^2 \kappa_1)^2 - k_o^4 \kappa_2^2 = 0$$

or

$$\beta^2 - k_o^2 \kappa_1 = \pm\, k_o^4 \kappa_2^2 \qquad (7.47)$$

Thus, we see that there are two solutions for β^2, which we will call β_1^2 for the + sign on the right-hand side of Equation (7.44), and β_2^2 for the − sign. Then,

$$\beta_1^2 = k_o^2 \kappa_1 + k_o^2 \kappa_2 = k_o^2 (\kappa_1 + \kappa_2) \qquad (7.48)$$

and

$$\beta_2{}^2 = k_0{}^2\kappa_1 - k_0{}^2\kappa_2 = k_0{}^2(\kappa_1 - \kappa_2) \qquad (7.49)$$

Plugging these values of β^2 back into Equation (7.46), one finds that for $\beta^2 = \beta_1{}^2$, $E_{oy} = j\,E_{ox}$, or $E_o = E_o\,(\mathbf{a}_x + j\,\mathbf{a}_y)$ which corresponds to a left-hand circularly polarized (LHCP) wave. Similarly, one can show that $\beta^2 = \beta_2{}^2$ corresponds to $E_{oy} = -j\,E_{ox}$ which implies it is a right hand circularly polarized (RHCP) wave.

Now, a linearly polarized wave can be equated to the sum of a RHCP wave and a left hand circularly polarized (LHCP) wave. For example, if the wave at $z = 0$ were polarized in the x direction it could be represented as

$$\mathbf{E} = E_o\,\mathbf{a}_x = (\mathbf{a}_x + j\,\mathbf{a}_y)\,(E_o/2) + (\mathbf{a}_x - j\,\mathbf{a}_y)\,(E_o/2)$$

After this wave propagates through the magnetized plasma from $z = 0$ to $z = \ell$ as shown in Figure 7.12, the electric field is given by

$$\mathbf{E} = (\mathbf{a}_x + j\,\mathbf{a}_y)\,(E_o/2)\,\exp\,(-j\,\beta_1\ell) + (\mathbf{a}_x - j\,\mathbf{a}_y)\,(E_o/2)\,\exp\,(-j\,\beta_2\ell)$$

$$= E_o\,\exp\,(-j\,[\beta_1 + \beta_2]\ell/2)\,\{\mathbf{a}_x\,\cos\,[\beta_1 - \beta_2]\ell/2 - \mathbf{a}_y\,\sin\,[\beta_1 - \beta_2]\ell/2\}$$

Thus, it is seen that as a result of the LHCP and RHCP wave components propagating at different speeds through the magnetized plasma, the wave is no longer polarized in the $\mathbf{a}_x$ direction. The electric field now makes an angle (the Faraday angle ϕ_F) with the x-axis, which was the electric field direction at $z = 0$. This situation is pictured in Figure 7.13.

The Faraday angle, ϕ_F, is given by

$$\phi_F = \tan^{-1}(E_y/E_x) = \tan^{-1}(-\tan\,[\beta_1 - \beta_2]\ell/2) = [\beta_2 - \beta_1]\ell/2 \quad (7.50)$$

From Equations (7.48) and (7.49) we choose

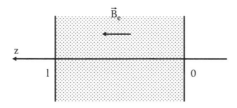

FIGURE 7.12 Magnetized plasma layer extending from z = 0 to z = 1 (e.g., the ionosphere).

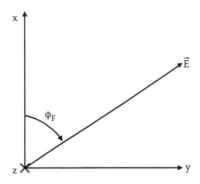

FIGURE 7.13 Geometry showing Faraday rotation to the final direction of polarization. The wave was initially polarized with an electric field along the *x*-axis.

$$\beta_1 = k_o (\kappa_1 + \kappa_2)^{1/2} \quad\quad (7.51)$$

and

$$\beta_2 = k_o (\kappa_1 - \kappa_2)^{1/2} \quad\quad (7.52)$$

Substituting Equations (7.51) and (7.52) into Equation (7.50) and using the inequality $(\kappa_2 / \kappa_1) \ll 1$, yields the following expression for the Faraday rotation angle:

$$\phi_F = (k_o \ell/2) (\kappa_2/\kappa_1^{1/2}) \quad\quad (7.53)$$

Expressions for κ_1 and κ_2 are given by Equations (7.37) and (7.38). If we take these to the high-frequency limit (i.e., $\omega_p^2 \ll \omega^2$ and $\omega_c^2 \ll \omega^2$), these expressions reduce to

$$\kappa_1 = 1$$

and

$$\kappa_2 = \omega_p^2 \omega_c / \omega^3$$

Then, substitution into Equation (7.50) gives

$$\phi_F = \omega_p^2 \omega_c \ell/(2 \, c \, \omega^2) \quad\quad (7.54)$$

Using the definitions of plasma frequency and electron cyclotron frequency in Equations (7.16) and (7.39), respectively, one may express the Faraday rotation angle in terms of the constant magnetic flux density, the electron density, and the path length in a uniform plasma as

$$\phi_F = (2.36 \times 10^4 \, B_e \, n_e \, \ell)/f^2 \qquad (7.55)$$

where mks units are used.

When the electron density is not uniform along the wave path it is more appropriate to express Equation (7.52) as

$$\phi_F = (2.36 \times 10^4 \, B_e \, n_T)/f^2 \qquad (7.56)$$

As an example, consider a total electron content $n_T = 10^{18} \, m^{-2}$ with $B_e = 7 \times 10^{-5}$ Tesla and $f = 10^9$ Hz. Then, Equation (7.53) gives a Faraday rotation angle of $\phi_F = 1.65$ radians or 95°. As a result of unpredictable variation in the ionospheric plasma, Faraday rotation may be significant and also unpredictable. To avoid an unacceptably large polarization mismatch loss, SATCOM signals are often circularly polarized rather than linearly polarized.

Problems

7.1. A transmitting antenna is located 50 m above the surface of the earth.

    a. Calculate the effective distance to the horizon assuming a smooth earth and "median refractivity" (i.e., $k_e = 4/3$).

    b. Recalculate the effective distance to the horizon under "super refractive" conditions with $k_e = 1.6$.

7.2. A radio wave is launched at sea level with an elevation angle of 1°. The refractive index is constant at 300 N-units up to an altitude of 100 m where it abruptly decreases to 200 n-units. Using Bouger's law, calculate the new elevation angle.

7.3. If the refractive index decreases continuously with increasing altitude by 75 N-units per kilometer, calculate the "effective earth radius" (i.e., calculate $k_e R_e$).

7.4. When plasma density in the ionosphere has a peak value of 10^{12} per cubic meter at an altitude of 350 km, what is the maximum usable frequency and the maximum range for a communications link using a single ionospheric reflection?

7.5. Assuming that the variation of electron density with height is as

$$N(h) = N_o \exp [-K (h - h_i)], \text{ for } h > h_i$$

and

$$N(h) = 0, \text{ for } h < h_i$$

Show that the total electron content $N_T = N_o/K$.

Evaluate N_T for $N_o = 10^{13}$ per cubic meter and $K^{-1} = 10$ km.

7.6. For the ionospheric model in Problem 7.5, calculate the dispersion if the center wave frequency is 12.4 GHz.

7.7. For the ionospheric model in Problem 7.5, calculate the differential delay at the opposite extremes of the transmitted frequency spectrum for a wave with center frequency 12.4 GHz and with a bandwidth of 400 MHz.

7.8. For the ionospheric model in Problem 7.5, calculate the maximum value of the Faraday rotation angle for a center frequency of 12.4 GHz. Repeat the calculation for a frequency of 1.24 GHz.

Bibliography

1. J.S. Seybold, *Introduction to RF Propagation* (John Wiley & Sons, Hoboken, NJ, 2005), 111–121.
2. [ITU, 453] ITU-R Recommendations, p. 453-6, "The Radio Refractive Index: Its Formula and Refractivity Data," Geneva, 1997.
3. [ITU, 530] ITU-R Recommendations, p. 530-7, "Propagation Data and Prediction Methods Required for Terrestrial Line-of-Sight Systems," Geneva, 1997.
4. [ITU, 834] ITU-R Recommendations, p. 834-2, "Effects of Tropospheric Refraction on Radiowave Propagation," Geneva, 1997.
5. R.E. Collin, *Antennas and Radiowave Propagation*. 2nd ed. (McGraw-Hill, New York, 1985), 388–400.
6. S. Ramo, J.R. Whinnery, and T. Van Duzer, *Fields and Waves in Communication Electronics*. 2nd ed. (John Wiley & Sons, New York, 1984), 703–715.

CHAPTER 8

Satellite Communications (SATCOM)

As described in Chapter 7, long-range communications by reflection of radiowaves from the ionosphere provide communications on a worldwide scale but are limited to frequencies below ~60 MHz. Thus, the bandwidth and the information capacity are severely limited. While shortwave radio broadcasts can be accommodated, television programming would require much more bandwidth.

This limitation has been overcome by placing artificial satellites of the earth in place far above the ionosphere. They are used as reflectors or relay stations in worldwide wireless communication systems that typically operate at frequencies one thousand times larger than the frequencies used in ionospheric reflection systems. We begin our consideration of satellite communications (SATCOM) by considering the fundamental principles involved in placing a satellite of the earth in geosynchronous orbit (i.e., an orbit in which the satellite position is stationary with respect to any location on the surface of the earth).

8.1 Satellite Fundamentals

8.1.1 Geostationary Earth Orbit (GEO)

The gravitational attractive force between two bodies of masses m_1 and m_2 separated by a distance R between their individual centers of mass is given by

$$F_G = g \frac{m_1 m_2}{R^2} \tag{8.1}$$

where the universal gravitational constant is g = 6.6726 × 10⁻¹¹ Newton meter²/kilogram².

As depicted in Figure 8.1, consider the circular orbit of a satellite of mass, m_s, around the earth of mass M_e = 6.03 × 10²⁴ kilograms. The satellite moves at a constant speed, v, in its orbit at a distance, R_a, from the center of the earth.

Due to its circular motion, the satellite experiences an outward centripetal force equal to

$$F_c = m_s \, \omega_s^2 \, R_a \tag{8.2}$$

where the angular speed $\omega_s = v_s/R_a$.

For an orbit in equilibrium, $F_c = F_G$, or from Equations (8.1) and (8.2),

$$m_s \omega_s^2 R_a = g \frac{m_s M_e}{R_a^2}$$

or

$$\omega_s = R_a^{-1.5} \sqrt{\mu_\oplus} \tag{8.3}$$

where $\mu_\oplus = gM_e = 4.02 \times 10^{14} \, N.m^2/kg$ is the earth's gravitational parameter.

The time required for the satellite to complete one orbit of the earth is $\tau_e = 2 \, \pi/\omega$. Then, using Equation (8.3),

$$\tau_e = \frac{2\pi R_a^{1.5}}{\sqrt{\mu_\oplus}} \tag{8.4}$$

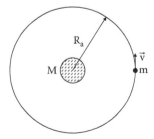

FIGURE 8.1   Circular orbit of a satellite of the earth.

and

$$v_s = \frac{2\pi R_a}{\tau_e} = \sqrt{\frac{\mu_\oplus}{R_a}} \qquad (8.5)$$

For a satellite in geosynchronous orbit, the time required for an orbit is 1 day, or

$$\tau_{synch} = 1\text{day} = 24 \times 3600\,\text{sec}$$

The distance of a geosynchronous satellite from the center of the earth is obtainable from Equations (8.4) and (8.6) as

$$R_{synch}^{1.5} = \frac{\sqrt{\mu_\oplus}}{2\pi} \tau_{synch} = \frac{2\times10^7}{2\pi} \times 2.4 \times 3.6 \times 10^4 = 2.75 \times 10^{11}\,\text{m}^{1.5}$$

or

$$R_{synch} = 42,300\,\text{km} \qquad (8.6)$$

The shortest distance from a geosynchronous satellite to the surface of the earth at sea level can be found by subtracting the earth radius from this value of 42,300 km (viz., 42,300 km − 6375 km = 35,900 km).

Geosynchronous satellites whose orbits coincide with the equatorial plane of the earth are widely used in SATCOM because antennas on the earth do not have to be re-aimed as time passes. The distance to such satellites from a given point on earth is constant.

8.1.2 Example of a GEO SATCOM System

The Intelsat system is based on four satellites in geosynchronous orbit around a plane passing through the earth's equator. These four satellites provide coverage of almost all the earth except for regions near the north and south poles. Two frequency bands are used, 1.6 GHz and 2.5 GHz.

The free space path loss for Intelsat is

$$L_f(dB) = 3.24 + 20\log R_{km} + 20\log f_{MHz}$$

$$= 32.4 + 20\log 36000 + 20\log 2500$$

$$= 193dB$$

Although this value may seem large, it is larger still for GEO SATCOM systems that operate at higher frequency as many of them do. The total path loss for Intelsat is even larger than the free space path loss because of excess attenuation caused by atmospheric gases and rain in the earth's atmosphere.

Fortunately, some of the practical restrictions that limited mobile phone systems do not apply to SATCOM. In SATCOM, transmitting antennas are not held close to a person's ear; thus, radiation health concerns, which limit power in a cell phone transmitter to 100 milliwatts, pose no limitation for SATCOM. Also, cell phone antennas were rather low gain, because they had to be both omnidirectional and easily portable. SATCOM antennas have a much higher gain, because they can be both relatively large and directed along a known transmitter-to-receiver path.

8.2 SATCOM Signal Attenuation

8.2.1 Attenuation Due to Atmospheric Gases

Attenuation of SATCOM signals by atmospheric gases is of concern at higher signal frequencies, in the range where wavelength is less than 2 cm. The molecular species most responsible for attenuation are molecular oxygen, O_2, and water vapor, H_2O.

Each of these molecules has its own set of frequencies where certain intermolecular resonances are excited, and where the absorption of radio waves is especially strong.

For SATCOM, the most important frequencies for resonant absorption are at 22 GHz in water vapor and at 60 GHz in molecular oxygen. The absorption at 60 GHz is especially strong, and communication system frequencies are usually chosen to avoid frequencies in the vicinity of 60 GHz.

Excess attenuation due to atmospheric gases along a zenith path (elevation angle equal to 90°) is plotted in Figure 8.2. What is plotted is total excess attenuation from sea level to the outer edge of the atmosphere. Two curves are shown, one for dry air and one with water vapor present. On the second

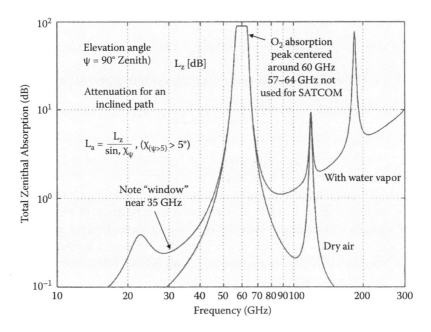

FIGURE 8.2 Excess attenuation of microwaves on a zenith path by atmospheric gases.

curve, note the presence of a propagation "window" around 35 GHz where attenuation is relatively low. This "window," located between the resonant absorption frequencies of 22 GHz and 60 GHz, is a favored choice for operation of many millimeter wave radar systems.

As stated, Figure 8.2 shows attenuation for a zenith path, L_z, where elevation angle $\psi = 90°$. For an inclined path ($\psi < 90°$), the total attenuation in atmospheric gases, L_G, may be simply found from the zenith attenuation as

$$L_G = L_z/\sin \psi \qquad (8.7)$$

where the range of elevation angle is limited to $\psi > 5°$.

8.2.2 Attenuation Due to Rain

The physical mechanism, which will be most responsible for attenuation of a radio frequency (RF) wave by rain, depends on the diameter of the rain droplet, D_R, compared with the wavelength. When the droplet is relatively small ($D_R \ll \lambda$), attenuation is primarily due to absorption of wave energy by the droplets. This phenomenon is known as Rayleigh attenuation. The attenuation cross section of a droplet has wavelength dependence as given by

$$C(D_R) \sim \frac{D_R{}^3}{\lambda} \tag{8.8}$$

The mean diameter of a droplet increases with the rain rate as

$$D_{mean} = 0.122 R_{rain}{}^{0.21} \tag{8.9}$$

where the droplet diameter is in millimeters, and the rain rate, R_{rain}, is in units of millimeters per hour.

Equation (8.8) indicates that at a given value of rain rate, attenuation will be linearly proportional to frequency. Equation (8.9) implies stronger dependence on the rain rate. These dependences are exhibited in Figure 8.3, where specific excess attenuation due to rain in dB/km is plotted as a function of frequency with rain rate as a parameter.

On the right side of Figure 8.3, the specific attenuation is seen to approach a constant value as frequency becomes very large (the optical limit). A different attenuation mechanism is dominant here (viz., scattering of the radio waves by droplets with diameters on the order of, or larger than, the wavelength). This process is known as Mie scattering.

As an alternative to using a graphical presentation of data, it is common to calculate the specific attenuation due to rain from the empirical expression

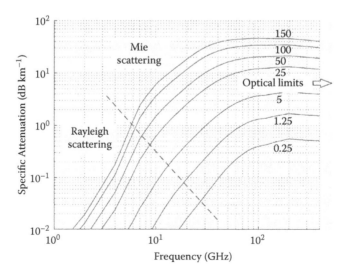

FIGURE 8.3 Specific rain attenuation versus frequency. Parameter is rain rate R in mm/hr.

$$\gamma_R = aR_{rain}^{b}(dB/km) \tag{8.10}$$

where the coefficients a and b are functions of signal frequency as specified in Table 8.1.

The total excess attenuation, due to rain, is then found by multiplying the specific attenuation by the path length through the rain, r_R:

$$L_r = r_R \gamma_R \tag{8.11}$$

The path length through the rain extends from the earth-station antenna up to an altitude where the temperature of the air is $0°$ centigrade (the frozen layer) as pictured in Figure 8.4. All heights in this figure are measured with respect to sea level. The rainy path length depends on the height of the frozen layer, h_R, the height of the antenna, h_A, and the elevation angle, ψ, as given by

$$r_R = (h_R - h_A)/\sin\psi \tag{8.12}$$

In Equation (8.12), the elevation angle is assumed to be greater than $5°$.

The height of the frozen layer will depend on the latitude of the base station. Values of h_R are displayed in Table 8.2.

TABLE 8.1   Parameters for Empirical Calculation of Specific Rain Attenuation (for Use with Equation 8.10)

F (GHz)	a	b
1	0.0000387	0.912
10	0.0101	1.276
20	0.0751	1.099
30	0.187	1.021
40	0.350	0.939

Source: From [ITU, 838] ITU-R, Recommendations, p. 838-1, "Specific Attenuation Model for Rain for Use in Prediction Methods," Geneva, 1992. With permission.

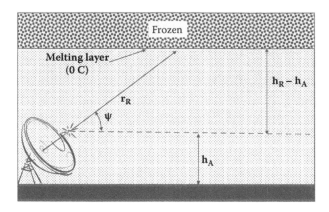

FIGURE 8.4 Geometry of rain attenuation path.

TABLE 8.2 Frozen Layer Height at Various Latitudes. Note that latitude in the southern hemisphere is negative. (from ITU, 618)

Latitude, χ_L (degrees)	Hemisphere	Altitude of frozen layer, h_R (kilometers)
$\chi_L > 23°$	Northern	$5 - 0.075 \, (\chi_L - 23°)$
$0° < \chi_L < 23°$	Northern	5
$-21° < \chi_L < 0°$	Southern	5
$-71° < \chi_L < -21°$	Southern	$5 + 0.1 \, (\chi_L + 21°)$
$\chi_L < -71°$	Southern	0

8.2.3 The Rain Rate Used in SATCOM System Design

SATCOM systems are usually designed in a conservative manner. They are required to operate successfully in heavy rainfall up to the rain rate that is exceeded only 0.01% of the time (about 1 hour per year). Such a rain rate is denoted by the symbol $R_{0.01}$, and the corresponding excess path loss due to this rain rate may be calculated using Equations (8.10) and (8.11), which is given by

$$L_{0.01} = a R_{0.01}^b r_R \tag{8.13}$$

where $L_{0.01}$ is in dB, $R_{0.01}$ is in mm/hr, and r_R is in km.

For elevation angles significantly smaller than 90°, Equation (8.13) may give an overestimate of the rainy path loss, because rain cells have a finite horizontal extent and may not fill the region between the antenna and the point where the ray contacts the frozen layer. This may be accounted for by multiplying the right-hand side of Equation (8.13) by a reduction factor that is somewhat smaller than one. In this book, we will take a more conservative design approach by assuming that the reduction factor always equals one.

The value of $R_{0.01}$ will depend upon geographical location. In Figure 8.5, a map of the United States is overlaid with regions of expected values of $R_{0.01}$. It may be seen that values of $R_{0.01}$ vary widely in the continental United States from 19 mm/hr to 98 mm/hr. Maps like this are also available for other parts of the world.

If a percentage of outage time other than 0.01% is chosen, the path loss can be calculated for the new percentage, p, using the following equation:

$$L_p = L_{0.01} \times 0.12 \, p^{-(0.546 + 0.043 \log p)} \tag{8.14}$$

Sample Problem:

A 30 GHz SATCOM system is designed to operate with rain rate as heavy as $R_{0.01}$ = 30 mm/hr with height of the frozen layer h_R = 5 km. The base station antenna is located 1 km above sea level and has an elevation angle of 60°.

Find the excess path loss due to rain.

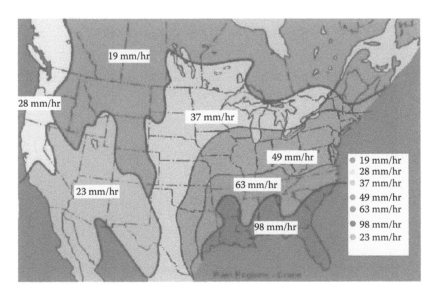

FIGURE 8.5   Rain rate map for the United States, $R_{0.01}$(mm/hr).

Solution:

From Table 8.1, we find that for 30 GHz, $a = 0.187$ and $b = 1.021$.

Then, the rainy path loss is found from Equations (8.13) and (8.12) as

$$L_{0.01} = aR_{0.01}^b \frac{h_R - h_A}{\sin \psi} = 0.187 \times 30^{1.021} \times (5-1)/\sin 60°$$

$$= 27.8 \text{ dB}$$

Note that when a SATCOM system is designed to operate in heavy rain, signal attenuation by atmospheric gases is negligible. Also, as we will see in the next section, the antenna noise temperature is dominated by the rain.

8.3 Design of GEO SATCOM Systems

8.3.1 Noise Calculations for SATCOM

The equivalent noise power at a receiver input assuming a lossless cable connecting the receiving antenna to the amplifier input was given in Chapter 2, Equation (2.16) as

$$P_N = k_B \Delta f \left[(1-\eta)T_0 + \eta\, T_A + (F-1)T_0 \right] \qquad (8.15)$$

For large receiving antennas such as those used in SATCOM ground stations, antenna Ohmic efficiency is close to 100%, and Equation (8.15) reduces to

$$P_N = k_B \Delta f \left[T_A + (F-1)T_0 \right] \qquad (8.16)$$

Because the noise factor, F, is usually specified for a given receiver, the unknown to be determined on the right-hand side of Equation (8.16) is antenna temperature, T_A.

Recall from Chapter 2 that noise temperature of a receiving antenna can in general have contributions from a number of sources:

$$T_A = f\,(T_s, T_b, T_{cosmic}, T_{atmospheric}) \qquad (8.17)$$

The first term in the brackets on the right-hand side of Equation (8.17) is the temperature of objects such as buildings, foliage, people, automobiles, and so forth, that may be positioned to lie in the beam of the receiving antenna. In SATCOM, the receiving antenna is usually aimed to avoid such objects, and so T_s may be considered to be negligible.

The second term is the "big bang" noise, which, while scientifically interesting, is too small (2.6°K) to contribute significantly to the noise temperature of the receiving antenna in a SATCOM system.

The third term, the cosmic noise, can have contributions from the galactic noise and from the sun or the moon, if they are in the bandwidth of the receiving antenna. The galactic noise is $\sim f^{-2.3}$, and will be negligible at the microwave and millimeter wave frequencies used in SATCOM. On the other hand, because the earth is rotating on its axis, and its axis is precessing with the seasons, the sun or the moon will occasionally come into view.

The cosmic noise sources are plotted as a function of frequency in Figure 8.6. In this figure, the moon and especially the sun are seen to have a substantial noise temperature in the microwave and millimeter-wave frequency bands.

As described in Chapter 2, when calculating the contribution of the sun or the moon to antenna temperature, it is appropriate to multiply their temperature by the fraction of the beam area they occupy. This fraction is given by

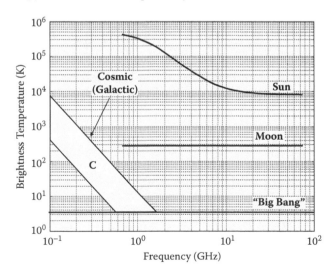

FIGURE 8.6 Cosmic noise temperatures vs frequency.

$$u = (0.5°/\text{beamwidth})^2 \qquad (8.18)$$

for beamwidth >0.50

Equation (8.18) reflects the fact that both the sun and the moon as seen on earth occupy a $0.5°$ angular diameter.

Finally, the fourth term on the right-hand side of Equation (8.17) is the noise contribution from atmospheric gases and from rain. Noise temperature of atmospheric gases was plotted in Figure 2.7. Rain temperature is usually taken as $260°K$.

Usually, either the rain or the atmospheric gases dominate the path loss through the atmosphere.

The preceding discussion may be summarized in the following expression for receiver antenna temperature in a SATCOM system. In general, with excess path loss in the atmosphere, L_{atmos}, the receiver antenna temperature will be

$$T_A = T_{atmos}(1 - 10^{-L_{atmos}/10}) + uT_{sun,moon} \times 10^{-L_{atmos}/10} \qquad (8.19)$$

If the rain is the dominant factor causing attenuation in the atmosphere, Equation (8.19) becomes

$$T_A = T_{rain}(1 - 10^{-L_{rain}/10}) + uT_{sun,moon} \times 10^{-L_{rain}/10} \qquad (8.20)$$

If gases dominate attenuation in the atmosphere, Equation (8.19) becomes

$$T_A = T_{gases}(1 - 10^{-L_{gases}/10}) + uT_{sun,moon} \times 10^{-L_{gases}/10} \qquad (8.21)$$

Finally, we note that if there is a lossy feeder cable between the receiving antenna and the receiver amplifier input, then Equation (8.16) is replaced by

$$P_N = k_B \Delta f T_{sys} \qquad (8.22)$$

and

$$T_{sys} = T_A \times 10^{-L_f/10} + T_f(1 - 10^{-L_f/10}) + (F - 1)T_0 \qquad (8.23)$$

where T_f is the feeder cable temperature, and L_f is the feeder cable loss.

Sample Calculation of Noise in SATCOM

Consider a SATCOM system operating at 30 GHz. The earth-based receiving antenna has an elevation angle near 90° and a beamwidth of 1°. The noise figure of the receiver is $F(dB) = 1$ dB (noise factor $F = 1.26$), and the feeder cable between the antenna and the receiver amplifier has negligible loss ($L_f \sim 0$). We will perform two calculations of equivalent noise input into the receiver: one calculation for the case of heavy rain ($R_{0.01} = 30$ mm/hr; height of the rain layer $h_R = 5$ km; receiving antenna height $h_A = 0$), and the second calculation without rain but with the receiving antenna looking directly at the sun. It will be instructive to see which situation gives the worst signal-to-noise ratio and thus should be used in conservative worst-case SATCOM system designs.

a. The Case of Heavy Rain

The excess path loss through the rain is calculated using Equation (8.13) and Table 8.1 as

$$L_{rain} = L_{0.01}$$

$$= aR_{0.01}^b r_R$$

$$= 0.187 \times 30^{1.021} \times 5$$

$$= 30.1 dB$$

The antenna temperature is calculated from Equation (8.20) as

$$T_A = T_{rain}(1 - 10^{-L_{rain}/10}) + uT_{sun,moon} \times 10^{-L_{rain}/10}$$

$$= 260(0.999) + \left(\frac{0.5^0}{1^0}\right)^2 10^4 \times 9.77 \times 10^{-4}$$

$$= 262^0 K$$

The equivalent noise input power is then calculated from Equation (8.16) as

$$P_{ni} = k_B \Delta f \left[T_A + (F-1)T_0\right]$$

$$= k_B \Delta f [262 + 0.26 \times 290]$$

$$= k_B \Delta f [337^0 K]$$

b. The Case of No Rain but Looking Directly at the Sun
Without rain, the excess path loss comes from atmospheric
gases.

From Figure 8.2, we find that at $f = 30$ GHz,

$$L_{gases} \approx 0.25 dB$$

We may also determine the noise temperature at $f = 30$ GHz
and zenith elevation by reference to Figure 2.7 as

$$T_{gases} \approx 25°K$$

Then, taking the noise temperature of the sun as 10,000°K,
the antenna temperature is found from Equation (8.19) as

$$T_A = T_{Gases}(1 - 10^{-0.25/10}) + uT_{sun} \times 10^{-0.25/10}$$

$$= 25^0 K(1 - 0.994) + \frac{10^4}{4} \times 0.994$$

$$= 2485.2^0 K$$

The equivalent noise input power is then calculated as

$$P_{ni} = k_B \Delta f \left[T_A + (F - 1)T_0 \right]$$

$$= k_B \Delta f [2560.6^0 K]$$

Comparing this answer with the answer to Part (a), we see
that in clear weather looking directly at the sun, the equivalent noise input power is 7.6 dB larger than in heavy rain.

However, the excess path loss is 29.8 dB larger in the
rain, and excess loss will diminish the strength of the signal.
Thus, we see that the signal-to-noise ratio in the presence
of heavy rain is about 20 dB smaller than in clear weather,
even when the receiving antenna is looking directly at the
sun. Therefore, the most conservative approach is to design
for operation in heavy rain.

8.3.2 Design of GEO SATCOM System for Wideband Transmission

A basic relationship for any communication system is

$$P_{s,min} = \gamma_{min} P_N = \frac{P_t \, G_t \, G_r}{L_t \, L_r \, L}$$

where γ_{min} is the minimum acceptable value of the signal-to-noise ratio.

In cell phone systems, we typically used this equation to find a maximum value of the path loss ($L = L_{max}$). In the link from the mobile unit to the base station, transmitter power, P_t, is limited by radiation safety considerations, and the gain of the transmitting antenna, G_t, is limited by requirements of omnidirectionality and portability. The base station antenna gain is also limited by the requirements of covering the entire cell. The feeder cable losses L_t and L_r are made as small as possible. Then, using a physical model of path loss, which relates L to the range between the transmitter and receiver, we could determine the maximum allowable value of cell radius.

In geosynchronous orbit (GSO) SATCOM, the range is fixed at ~36,000 km. However, other parameters such as the size of the receiving antenna can be chosen to match the desired system performance. Consider a SATCOM system intended for high data rate transmission, such as for example would be required to handle TV programming. This would require a bandwidth on the order of Δf=10 MHz.

If the center frequency was f = 30 GHz, and the effective area of the antenna on the satellite was 1 m², the gain of this antenna would be G_t = 51 dBi. (See Equation 3.77.) At the specified values of f and Δf, the most power commercially available from a single source (helix traveling wave tube [TWT] vacuum tube) is approximately P_t = 30 dBW. Assuming a feeder cable loss on the satellite of 1 dB, the effective isotropic radiated power is EIRP = 80 dBW.

Let us calculate the size of receiving antenna required, assuming the receiver is located in Baltimore, Maryland, which has a latitude of 39 degrees north (ϕ_{lat} = 39°), an elevation of 10 m, and a rain rate exceeded only 0.01% of the time of $R_{0.01}$ = 39 mm/ hr. We will also specify an antenna elevation angle of ψ = 80°, a receiver noise figure F(dB) = 3dB, a minimum signal-to-noise ratio of γ_{min} = 6 dB, and a receiver feeder cable loss of 2 dB.

We design the SATCOM system for operation up to the $R_{0.01}$ rain rate. The path loss from the satellite to the receiving antenna has contributions from both the free space path loss and the rain excess loss as

$$L(dB) = L_F(dB) + L_{rain}(dB)$$

The free space path loss is calculated as

$$L_F(dB) = 32.4 + 20\log R_{km} + 20\log f_{MHz}$$

$$= 32.4 + 20\log 36000 + 20\log 30000$$

$$= 213dB$$

The excess loss due to rain is calculated as

$$L_{rain}(dB) = L_{0.01}(dB)$$

$$= aR_{0.01}^b \gamma_R$$

$$= 0.187 \times 39^{1.021} \times \left[5 - 0.075\left(\phi_{lat} - 23^0\right) - 0.01\right] / \sin 80^0$$

$$= 30.4dB$$

Then, the total path loss is

$$L(dB) = L_F(dB) + L_{rain}(dB)$$

$$= 213 + 30.4$$

$$= 243dB$$

The equivalent noise input power is

$$P_N = k_B \Delta f T_{sys}$$

$$= k_B \Delta f \left[T_A 10^{-L_r/10} + T_f(1 - 10^{-L_f/10}) + (F-1)T_0\right]$$

$$= 1.38 \times 10^{-23} \times 10^7 \left[260 \times 10^{-0.2} + 290(1 - 10^{-0.2}) + 290\right]$$

$$= 7.73x10^{-14} Watts$$

or

$$P_N(dBW) = -131 \text{ dBW}$$

So that the minimum required input signal power is

$$P_{s,\min}(dBW) = \gamma_{MIN} + P_N(dBW)$$

$$= 6 - 131$$

$$= -125 dBW$$

Next, the minimum receiving antenna gain is calculated as

$$G_{r,\min}(dBi) = P_{s,\min}(dBW) + L_r(dB) + L(dB)$$

$$+ L_t(dB) - G_t(dBi) - P_t(dBW)$$

$$= -125 + 2 + 243 + 1 - 51 - 30$$

$$= 40 dBi$$

or

$$G_{r,\min} = 10^4$$

The corresponding effective area of the receiving antenna would be

$$A_{er,\min} = \frac{G_{r,\min}\lambda^2}{4\pi}$$

$$= \frac{10^4}{4\pi} cm^2$$

$$= 800 cm^2$$

Assuming the receiving antenna was a parabolic reflector, its effective area would be 55% of its physical area. Then, the physical area of the receiving antenna would be $A_r = 0.145$ m², and the diameter would be 0.43 m.

Finally, we note that any attempt to decrease the size of the receiving antenna by making the gain of the transmitting antenna larger will have an offsetting disadvantage (viz., the illuminated area on the earth will become smaller). For a transmitting antenna gain of $G_t = 10^5$, the beamwidth may be estimated using Equation (1.11) and assuming $BW_\varphi = BW_\theta$. This gives

Beamwidth $\sim [41{,}000/10^5]^{1/2}$ degrees $= 0.64$ degrees

Then, the area illuminated on the earth is $(\pi/4)(36000 \sin 0.64°)^2$ km$^2 \sim 10^5$ km^2.

For some systems such as satellite-based TV programming transmission, which are aimed at broad area coverage, this value of illuminated area may be insufficient.

8.4 Medium Earth Orbit (MEO) Satellites

An MEO satellite is one whose time to orbit the earth varies from 2 to 12 hr. The first communications satellite, Telstar, was an MEO satellite. Today the most important MEO satellites are part of a global positioning system (GPS).

8.4.1 Global Positioning System (GPS)

GPS can determine the location of a transmitter on earth (e.g., in your automobile) with precision on the order of 1 m. This is accomplished by transmitting a pulsed signal to three different satellites, and precisely measuring the transmission time (and thus the transmission distance) between the earth-based GPS unit and each satellite. The location of the GPS unit on the surface of the earth can then be determined by using the well-known technique of triangulation as depicted in Figure 8.7. By employing a fourth satellite, the altitude of the earth-based GPS unit can also be determined. The civilian GPS frequency is 1.575 GHz. The satellites are closer to the earth than GEO satellites and complete an orbit of the earth in 12 hr. Using Equation (8.4), one can calculate that τ_e = 12 hr corresponds to an orbit with a distance from the center of the earth of R_a = 26,700 km.

To achieve an accuracy in position location of 1 m, clocks in the satellites and in the earth-based GPS unit have to be very accurate (atomic clocks are used) and have to be synchronized within a few nanoseconds because electromagnetic waves travel a distance of 1 m in 3 ns. It is interesting that the effect of gravity on clocks, an effect predicted by Albert Einstein's theory of general relativity, has to be taken into account if the required degree of clock synchronization is to be maintained.

8.4.2 General Relativity, Special Relativity, and the Synchronization of Clocks

The earth-based GPS unit is ~6400 km from the center of the earth, while GEO satellites are ~26,700 km from the center of the

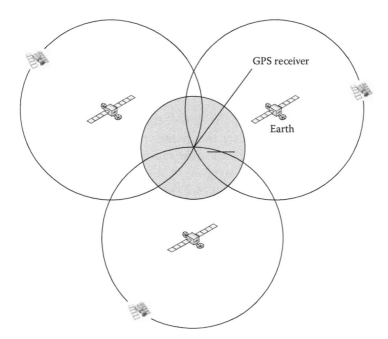

GPS receiver

Earth

FIGURE 8.7 Triangulation for GPS.

earth. Thus, as given by Equation (8.1), the gravitational force on the earth-based clock is ~17 times stronger than the gravitational force on the clocks in the satellites. According to general relativity, the stronger gravitational field on the earth-based clock will result in it running more slowly than the satellite-based clocks. By some measures, the difference is small but must be accounted for in the design of GPS systems. If an earthbound clock and satellite clocks are initially synchronized, then after one orbit of the satellites, the difference in the time indicated on the clocks would be large enough to produce an error in the location calculated of many kilometers. General relativity has found wide acceptance in astronomy and cosmology studies, but this is the first practical application of the theory of general relativity to an earth-based system.

Einstein's prediction for the effect of gravity on clocks can be derived in a straightforward way for atomic clocks. The total energy of an atom with mass m has two parts as follows:

1. Rest mass energy,

$$W_m = m\, c^2 \qquad (8.24a)$$

2. Gravitational potential energy,

$$W_G = -R_a \, F_G = -g \, m \, M_e/R_a \qquad (8.24b)$$

where we have used the expression for gravitational force, F_G, in Equation (8.1).

In Equation (8.24b), the sign of the gravitational energy is negative because the greater the distance between the earth and the atom, the larger will be the gravitational potential energy.

The total energy of an atom in the earth's gravitational field is then found from Equation (8.24) as

$$W = W_m + W_G = m \, c^2 - g \, m \, M_e/R_a = m \, c^2 [1 - \mu_\oplus/(c^2 \, R_a)] \qquad (8.25)$$

Thus, an atom in its ground state will have an energy that is decreased by the factor $[1 - \mu_\oplus/(c^2 \, R_a)]$ due to the effect of the earth's gravitational field. We assume that the energy of the excited states of the atom will be decreased by the same factor. Finally, we note that this implies that a photon emitted as a result of transition of the atom between excited states will have energy

$$h \, f' = hf \, [1 - \mu_\oplus/(c^2 \, R_a)] \qquad (8.26)$$

where f is the photon frequency in the absence of a gravitational field.

Thus, a clock based on the atomic transition will run more slowly in a gravitational field (i.e., it will produce lower-frequency photons). The elapsed time measured by a clock in a gravitational field is then

$$\Delta t' = \Delta t \, [1 - \mu_\oplus/(c^2 \, R_a)] \qquad (8.27)$$

where Δt is elapsed time measured by a gravity-free clock.

We can use Equation (8.27) to compare elapsed time measured on the surface of the earth with elapsed time measured by a clock on a GPS satellite. Supposing that the clocks are initially synchronized, we can calculate the time measured by each clock after a 12-hr orbit. The clock on the surface of the earth will measure an elapsed time of

$$(\Delta t')_e = 12 \text{ hours } [1 - \mu_\oplus/(c^2 \, R_{ae})] = 12 \text{ hours}$$

$$[1 - 4.02 \times 10^{14}/(9 \times 10^{16} \times 6.375 \times 10^6)]$$

$$= 12 \text{ hours} - 30 \text{ } \mu\text{sec}$$

The clock on the satellite will measure an elapsed time of

$$(\Delta t')_s = 12 \text{ hours} [1 - \mu_\oplus/(c^2 \text{ } R_{as})] = 12 \text{ hours}$$

$$[1 - 4.02 \times 10^{14}/9 \times 10^{16} \times 26.7 \times 10^6)$$

$$= 12 \text{ hours} - 7 \text{ } \mu\text{sec}$$

Thus, due to gravitational effects, the satellite clock will be faster than the earth-based clock by 23 μsec.

There is also a slowing effect on the satellite clock predicted by Einstein's theory of special relativity. For a GPS system, this slowing effect is considerably smaller than the gravitational effect described above, but it should be taken into account. The satellite is moving at a speed v_s relative to an earthbound clock and according to special relativity, the satellite clock will run slower by a factor of $[1 - (v_s/c)^2]$. For a GPS satellite at R_a = 26,700 km, we calculate using Equation (8.5) that v_s = 3.88 km/s so that $[1 - (v_s/c)^2]$ = 8.3 × 10^{-11}. Then, due to special relativity effects, during a 12-hr orbit the satellite clock will run slower by

$$12 \text{ hours} \times 8.3 \times 10^{-11} = 3.5 \text{ microseconds}$$

The net effect of general relativity effects due to gravity and special relativity effects due to relative speed is that the satellite clock will run faster than the earth-bound clock by 23 microseconds − 3.5 microseconds = 19.5 microseconds after a 12-hr orbit.

If a correction is not made for the relativistic effects, the distance from the GPS unit on the ground to a satellite will be calculated with an error of 19.5 × 10^{-6} × 3 × 10^8 m = 5.8 km.

8.5 Low Earth Orbit (LEO) Communication Satellites

Geosynchronous orbit (GSO) satellites, due to their large distance from the earth (36,000 km) are not well suited to two-way mobile telephone communications, because of the large path loss (implying the need for large antennas) and because of the significant propagation delay time. Low earth orbit (LEO) satellite systems overcome these limitations by having satellites much closer to the earth (e.g., 778 km); however, an LEO system is much more complex.

LEO satellites orbit the earth in a relatively short time, as may be seen from the following calculation. Using Equation (8.5) one can calculate the velocity of an LEO satellite at an altitude of 778 km:

$$v_s = \sqrt{\frac{\mu_\oplus}{R_a}}$$

$$= 2 \times 10^7/R_a^{1/2}$$

$$= 2 \times 10^7/[(778 + 6375) \times 10^3]^{1/2}$$

$$= 7.48 \text{ km/sec}$$

The time for the LEO satellite to orbit the earth is then calculated as

$$T = 2\pi R_a/v_s$$

$$= 2\pi (778 + 6375)/7.48$$

$$= 6000 \text{ seconds}$$

$$= 1.66 \text{ hours}$$

Thus, an LEO satellite will be moving quickly with respect to the surface of the earth. There will be a significant Doppler frequency shift in a communication system based on LEO satellites. Also, a large number of satellites will be required to maintain a continuous communication link.

8.5.1 The Iridium LEO SATCOM System

An example of an LEO system is Iridium, which became operational in November 1998. Iridium was initially owned by Motorola but went into Chapter 11 bankruptcy in less than 1 year, due to an insufficient customer base. Tariffs were so high that only a few large organizations like news networks found it worthwhile to subscribe. Iridium was acquired by a group of private investors who founded Iridium Satellite LLC; they paid $25 million for a system that had cost on the order of $6 billion, and they succeeded in reorganizing the company on a profitable basis. Services were reestablished in 2001, and by the end of 2006, there were approximately 170,000 subscribers. Rates from Iridium mobile units to land lines are ~$1.50 per minute, and the mobile units have an initial cost between \$1000 and \$2000. The system is used extensively by the U.S. Department of Defense and by the maritime, aviation, petroleum, and news reporting industries.

Iridium consists of 72 satellites in six equally spaced orbital planes. Thus, there are 12 satellites per orbital plane, six of which are operational plus one spare. The six orbital planes are sketched in Figure 8.8. The satellites are at an altitude of 778 km, and the corresponding round-trip propagation time for a signal is ~6 milliseconds. The center operating frequency is 1.625 GHz. Iridium provides voice and data communications using handheld units, and covers the whole earth including the poles. It is not dependent on the proximity of base stations and so can operate in undeveloped areas.

8.5.2 Path Loss in LEO SATCOM

It should be noted that the distance from a fixed point on earth to an Iridium satellite varies with time. The distance is 778 km when the satellite is directly overhead, but is 3244 km when the satellite first comes into view or when it vanishes beyond the horizon.

The situation is pictured in Figure 8.9, from which it should be clear that the distance depends on the elevation angle, ψ.

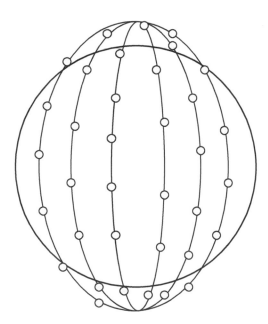

FIGURE 8.8 Iridium LEO SATCOM system (6 orbits, 12 satellites per orbit).

When the satellite first appears above the horizon ($\psi = 0$) as shown in the figure, the distance is given by

$$d = [(R_e + h_{sat})^2 - R_e^2]^{1/2}$$
$$= [(6375 + 778)^2 - 6375^2]^{1/2}$$
$$= 3244 \text{ km}$$

One can now calculate the free space path loss for the LEO communication system using Equation (5.6):

$$L_F(dB) = 32.4 + 20 \log R_{km} + 20 \log f_{MHz}$$

When the satellite is directly overhead, $R_{km} = 778$ and using $f_{MHz} = 1625$, one obtains a minimum value of the free space path loss

$$L_{min} = 32.4 + 20 \log 778 + 20 \log 1625$$
$$= 154 \text{ dB}$$

When the satellite is on the horizon, the range is 3244 km and the free space path loss has a maximum value

$$L_{max} = 32.4 + 20 \log 3244 + 20 \log 1625$$
$$= 167 \text{ dB}$$

Path loss as a function of elevation angle is plotted in Figure 8.10. Path loss as a function of time is plotted in Figure 8.11.

In addition, for small elevation angles, shadowing by buildings and hills must be taken into account. A margin of 13 dB is usually added to account for shadowing in a suburban mid-rise

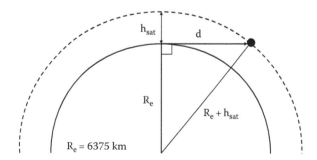

FIGURE 8.9 Geometry for determining the maximum distance to LEO satellite.

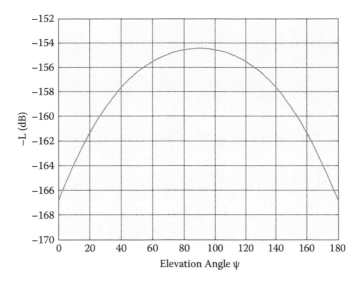

FIGURE 8.10 Path loss versus elevation angle, ψ, for LEO SATCOM system.

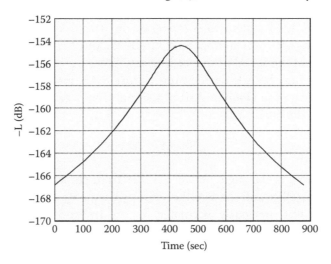

FIGURE 8.11 Path loss for Iridium as a function of time.

environment. For a downtown urban high-rise environment, 16 dB of margin is added. Finally, it should be noted that the satellite spends considerably more time at small values of elevation angle than near zenith. This is indicated by the plot in Figure 8.12 where a "p.d.f." of the elevation angle is plotted; of course the elevation angle ψ is not a random variable, but the plot may be used like a p.d.f. to calculate the fraction of time that ψ spends in a given range of values.

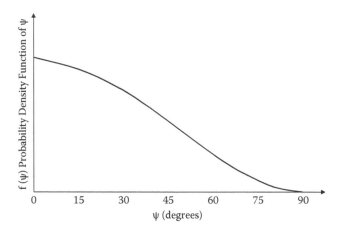

FIGURE 8.12 Probability density function of elevation angle ψ.

8.5.3 Doppler Shift in LEO SATCOM

Compensation is required at the earth-based receivers in an LEO SATCOM system to compensate for the frequency up-shift of the signal from approaching satellites, and the frequency down-shift of the signal from receding satellites. The maximum Doppler shift may be calculated using Figure 8.13. From this figure, we calculate the Doppler shift of a signal from a satellite as it appears over the horizon as

$$\delta f = f\,(v_s/c)\,\cos\xi \tag{8.28}$$

where ξ is the angle between a line from the earth-based unit to the satellite and the satellite velocity vector $\mathbf{v}_s$.

It was previously calculated for the Iridium satellites that $v_{s\,=}$ 7.48×10^3 m/s. One can also calculate from Figure 8.13 that when the satellite is at the horizon,

$$\cos\xi = R_e/(R_e + h_{sat})$$

$$= 6375/(6375 + 778)$$

$$= 0.892$$

Then, from Equation (8.28), the maximum Doppler shift is

$$\delta f\,(\mathrm{MHz}) = 1625 \times (7.48 \times 10^3/3 \times 10^8) \times 0.892$$

$$= 0.036\ \mathrm{MHz} = 36\ \mathrm{kHz}$$

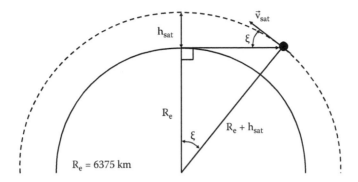

FIGURE 8.13 Geometry for calculating maximum Doppler shift in LEO SATCOM.

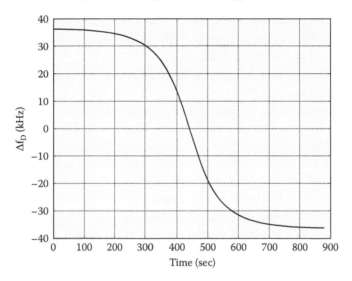

FIGURE 8.14 Doppler shift as a function of time (for Iridium).

We thus see a substantial Doppler shift when the satellite is at the horizon. The Doppler shift is of course zero when the satellite is directly overhead. The Doppler shift as a function of time is plotted in Figure 8.14.

Problems

8.1. Calculate the orbital velocity and the period of the following satellites of the earth:

a. The moon, which is 384,000 km from the earth's surface

b. A low earth orbit (LEO) communications satellite at an altitude of 650 km

8.2. For a GSO SATCOM system, compute the rain attenuation exceeded for only 0.001% of the time [i.e., calculate $R(0.001)$] in a region where $R(0.01)$ is 30 mm/hr and the latitude is approximately 40 degrees north. The elevation angle of the receiving antenna is 80°, and the signal frequency is 30 GHz.

8.3. For a GSO SATCOM system, compute the total atmospheric gas attenuation for an elevation angle of 80° and a signal frequency of 30 GHz.

8.4. A receiving antenna on earth, located at 55 degrees north latitude, has a 3 dB beamwidth of 1° at 10 GHz. Its elevation angle is 50°, and it is pointing directly at the moon. Rain is falling at a rate of 25 mm/hr. The feeder cable loss is 3 dB, and the receiver noise figure is 2 dB. Calculate the equivalent noise input power to the receiver.

8.5. Calculate the velocity of an LEO satellite at an altitude of 1000 km.

8.6. For the LEO satellite described in Problem 8.5, calculate the maximum Doppler shift if the center signal frequency is 15 GHz.

8.7. For the LEO satellite in Problem 8.5, estimate the maximum path loss.

8.8. For the LEO satellite described in Problem 8.5, calculate the maximum one-way propagation time.

8.9. If a clock is placed on the surface of the moon, how much of a time difference will there be compared with an earth-bound clock after 1 month due to gravitational effects? (The mass of the moon is 7.36×10^{22} kg, and the radius of the moon is 1737 km.)

Bibliography

1. [ITU, 838] ITU-R, Recommendations, p. 838-1, "Specific Attenuation Model for Rain for Use in Prediction Methods," Geneva, 1992.

2. [ITU-R, 618] ITU-R Recommendation p. 618-5, "Propagation Data and Prediction Methods Required for the Design of Earth-Space Telecommunication Systems," Geneva, 1997.

3. S. Hawkings and L. Mlodinow, *A Briefer History of Time* (Bantam Dell, New York, 2005), 48.

4. Wikipedia, the free encyclopedia, "Iridium satellite constellation." http://en.winkipedia.org/wiki/Iridium_satellite_constellation.

5. S.R. Saunders, *Antennas and Propagation for Wireless Communication Systems* (John Wiley & Sons, Chichester, UK, 1999), 126–135, 146–147.

6. N. Ashby, "Relativity in the Global Positioning System," *Physics Today* 55 (2002): 34–36.

Appendix A

TABLE OF PHYSICAL CONSTANTS

Description	Symbol	Numerical Value
Boltzmann constant	k_B	1.38×10^{-23} Joules/°K
Distance of a geosynchronous orbit (GSO) satellite from the center of the earth	R_{synch}	42,375 kilometers
Electron charge	e	1.6×10^{-19} Coulombs
Electron rest mass	m	9.11×10^{-31} kilograms
Gravitational constant	g	6.6726×10^{-11} Newton meter²/ kilogram²
Gravitational parameter of the earth	$\mu_\oplus$	4.02×10^{14} Newton meter²/ kilogram
Mass of the earth	M_e	6.03×10^{24} kilograms
Permeability of free space	μ_o	$4\pi \times 10^{-7}$ Henries/meter
Permittivity of free space	ε_o	$(36\pi \times 10^9)^{-1}$ Farads/meter
Planck's constant divided by 2π	$\hbar$	1.05443×10^{-34} Joule seconds
Radius of the earth	R_e	6375 kilometers
Speed of an electromagnetic wave in vacuum	c	2.998×10^8 meters/second
Wave impedance of free space	Z_o	120π Ohms

Appendix B

DEL OPERATORS IN CARTESIAN AND SPHERICAL COORDINATES

Operation	Cartesian Coordinates (x,y,z)	Spherical Coordinates (r,θ,φ)
Definition of coordinates	$\begin{cases} x = r\sin\theta\cos\phi \\ y = r\sin\theta\sin\phi \\ z = r\cos\theta \end{cases}$	$r = \left(x^2 + y^2 + z^2\right)^{-\frac{1}{2}}$ $\theta = \cos^{-1}(z/r)$ $\phi = \tan^{-1}(y/x)$
A	$A_x\hat{X} + A_y\hat{y} + A_z\hat{Z}$	$A_r\hat{r} + A_\theta\hat{\theta} + A_\phi\hat{\phi}$
∇f	$\dfrac{\partial f}{\partial x}\hat{x} + \dfrac{\partial f}{\partial y}\hat{y} + \dfrac{\partial f}{\partial z}\hat{z}$	$\dfrac{\partial f}{\partial r}\hat{r} + \dfrac{1}{r}\dfrac{\partial f}{\partial \theta}\hat{\theta} + \dfrac{1}{r\sin\theta}\dfrac{\partial f}{\partial \phi}\hat{\phi}$
$\nabla \cdot A$	$\dfrac{\partial A_x}{\partial x} + \dfrac{\partial A_y}{\partial y} + \dfrac{\partial A_z}{\partial z}$	$\dfrac{1}{r^2}\dfrac{\partial(r^2 A_r)}{\partial r} + \dfrac{1}{r\sin\theta\,\partial\theta}(A_\theta\sin\theta)$ $+ \dfrac{1}{r\sin\theta}\dfrac{\partial A_\phi}{\partial \phi}$
$\nabla \times A$	$\left(\dfrac{\partial A_z}{\partial y} - \dfrac{\partial A_y}{\partial z}\right)\hat{x}$ $+\left(\dfrac{\partial A_x}{\partial z} - \dfrac{\partial A_z}{\partial x}\right)\hat{y}$ $+\left(\dfrac{\partial A_y}{\partial x} - \dfrac{\partial A_z}{\partial y}\right)\hat{z}$	$\dfrac{1}{r\sin\theta}\left(\dfrac{\partial}{\partial\theta}(A\phi\sin\theta) - \dfrac{\partial A_\theta}{\partial\phi}\right)\hat{r}$ $+\dfrac{1}{r}\left(\dfrac{1}{\sin\theta}\dfrac{\partial A_r}{\partial\phi} - \dfrac{\partial}{\partial_r}(r\,A\phi)\right)\hat{\theta}$ $+\dfrac{1}{r}\left(\dfrac{\partial}{\partial_r}(r\,A\theta) - \dfrac{\partial A_r}{\partial\theta}\right)\hat{\phi}$

(Continued)

(CONTINUED) DEL OPERATORS IN CARTESIAN AND SPHERICAL COORDINATES

Operation	Cartesian Coordinates (x,y,z)	Spherical Coordinates (r,θ,φ)
$\Delta f = \nabla^2 f$	$\dfrac{\partial^2 f}{\partial x^2} + \dfrac{\partial^2 f}{\partial y^2} + \dfrac{\partial^2 f}{\partial z^2}$	$\dfrac{1}{r^2}\dfrac{\partial}{\partial r}\left(r^2\dfrac{\partial f}{\partial r}\right)$
		$+ \dfrac{1}{r^2\sin\theta}\dfrac{\partial}{\partial\theta}\left(\sin\theta\dfrac{\partial f}{\partial\theta}\right)$
		$+ \dfrac{1}{r^2\sin^2\theta}\dfrac{\partial^2 f}{\partial\phi^2}$
$\Delta A = \nabla^2 A$	$\Delta A_z \hat{x}$ $+ \Delta A_y \hat{y}$ $+ \Delta A_z \hat{z}$	$\left(\Delta A_r - \dfrac{2A_r}{r^2} - \dfrac{2A_\theta\cos\theta}{r^2\sin\theta} - \dfrac{2}{r^2}\dfrac{\partial A_\theta}{\partial\theta} - \dfrac{2}{r^2\sin\theta}\dfrac{\partial A_\phi}{\partial\phi}\right)\hat{r}$
		$\left(\Delta A_\theta - \dfrac{A_\theta}{r^2\sin^2\theta} - \dfrac{2}{r^2}\dfrac{\partial A_r}{\partial\theta} - \dfrac{2\cos\theta}{r^2\sin^2\theta}\dfrac{\partial A_\phi}{\partial\phi}\right)\hat{\theta}$
		$\left(\Delta A_\phi - \dfrac{A_\phi}{r^2\sin^2\theta} + \dfrac{2}{r^2\sin^2\theta}\dfrac{\partial A_r}{\partial\phi} + \dfrac{2\cos\theta}{r^2\sin^2\theta}\dfrac{\partial A_\theta}{\partial\phi}\right)\hat{\phi}$

Appendix C

DIFFERENTIAL LINE, AREA, AND VOLUME IN CARTESIAN AND SPHERICAL COORDINATES

Differential displacement	$dl = dx\hat{x} + dy\hat{y} + dz\hat{z}$	$dl = dr\hat{r} + rd\theta\hat{\theta} + r\sin\theta \, d\phi\hat{\phi}$
Area	$dS = dydz\hat{x}$	$dS = r^2\sin\theta \, d\theta d\phi\hat{r}$
	$+ \, dxdz\hat{y}$	$+ \, r\sin\theta \, d\theta d\phi\hat{\theta}$
	$+ \, dxdy\hat{z}$	$+ \, r \, drd\theta\hat{\phi}$
Differential volume	$dv = dxdydz$	$dv = r^2\sin\theta \, drd\theta d\phi$

Nomenclature

English Alphabet

A	Vector potential (Weber/meter, Wb/m)
A_e	Effective area of a receiving antenna
$A(\theta)$	Array pattern function for omnidirectional antenna
a	Dimension of a rectangular microstrip patch separating the radiating edges
a	Parameter used in calculating specific rain attenuation (Equation 8.10)
$\mathbf{a}_\theta, \mathbf{a}_\theta, \mathbf{a}_z$	Unit vectors in spherical coordinates
$\mathbf{a}_x, \mathbf{a}_y, \mathbf{a}_z$	Unit vectors in Cartesian coordinates
B	Magnetic flux density (Tesla)
B_e	Constant magnetic flux density due to the earth's magnetic field
BW	3-dB beamwidth
BW_ϕ	3 dB beamwidth in the direction of the azimuthal angle ϕ
BW_θ	3 dB beamwidth in the direction of the polar angle θ
b	Dimension of a rectangular microstrip patch along a radiating edge
b	Parameter used in calculating specific rain attenuation (Equation 8.10)
C	Capacitance (Farads)
C_I	Information capacity of a communication system (bits/sec)
C_1, C_2	Causebrook correction factors (Equation 5.30)
c	Speed of an electromagnetic wave in vacuum (~3 × 10⁸ meters/second)
D	Directivity of an antenna or of an array of antennas
D_1	Directivity of one antenna in an array of similar antennas
D_A	Array directivity factor [$D = D_1 D_A$]
D_{mean}	Mean diameter of rain droplet
D_R	Diameter of rain droplet
D_{slot}	Directivity of a slot antenna
d	Distance (e.g., from antenna to horizon)
d_B	Spacing between diffracting edges of buildings along communication path
d_H	Horizontal displacement between antennas
d_i	Distance between interfering co-channel cells

d_{max}	Maximum distance covered in ionospheric bounce communications
d_s	Distance between buildings on opposite sides of a street (width of the street)
ds	Incremental area with vector directed normal to the surface
d_V	Vertical displacement between antennas
E	Phasor representation of the electric field
E_d	Magnitude of the phasor representing the electric field of the signal in the dominant channel
E_i	Electric field in the incident wave
E_r	Electric field in the reflected wave
E_s	Magnitude of the phasor representing the signal electric field
E_t	Electric field in the transmitted wave
$E_\perp$	Electric field component perpendicular to the plane of incidence
$E_{||}$	Electric field component parallel to the plane of incidence
EIRP	Effective isotropic radiated power (Watts)
$E(t)$	Electric field (Volts/meter)
EXP	Expected value
e	Magnitude of the charge of an electron (1.6×10^{-19} Coulombs)
exp	Exponential
F	Force (Newtons)
F	Noise factor
F_c	Centripetal force
F_G	Gravitational force
$F(dB)$	Noise figure
f	Frequency (Hertz, Hz)
$\underline{f}$	Dimensionless frequency normalized to 1 GHz
f_{MHz}	Frequency in megahertz
f_{muf}	Maximum usable frequency in ionospheric communications (Equation 7.22)
f_p	Plasma frequency (Hertz, Hz) (Equation 7.17)
$f(\theta)$	Pattern function of a linear dipole antenna
$f_h(\theta)$	Pattern function of a half-wave dipole antenna
G	Antenna gain
G_r	Gain of receiving antenna
G_t	Gain of transmitting antenna
g	Universal gravitational constant
H	Phasor representation of the magnetic field

H(t)	Magnetic field (Amperes/meter, A./m.)
h$_r$	Unit vector indicating polarization that antenna is designed to receive
h$_t$	Unit vector indicating polarization of transmitted wave
h	Polarization unit vector
h	Half length of a dipole antenna
$\hbar$	Planck's constant divided by 2π (1.05443 × 10–34 Joule seconds)
h$_A$	Height of an antenna above sea level
h$_b$	Height of the base station antenna
h$_e$	Excess height of a knife edge with respect to the direct path
h$_i$	Height of the ionosphere (meters)
h$_m$	Height of the mobile antenna (e.g., the cell phone antenna)
h$_o$	Prevailing height of buildings
h$_{sat}$	Height of a satellite
h$_T$	Altitude at which there is a sudden transition in tropospheric refractive index
I	Current (Amperes, A)
I	Current phasor
I_i	Input current (e.g., at the central feedpoint of a linear dipole antenna)
I_n	Input current to the nth antenna in an array
I_o	Current amplitude
J$_s$	Surface current density (Amperes/meter)
J	Current density (Amperes/meter²)
j	Imaginary number equal to the square root of –1
K	Rice parameter (See Equation 6.51)
K_p	Polarization mismatch factor
k	Wavevector (meters^{-1})
k	Wavenumber; the magnitude of the wavevector
k_e	Effective earth radius factor
L	Path loss
$L(A)$	Excess path loss due to atmospheric gases and rain
L_B	Excess path loss due to penetrating the outer walls of a building
L_{BS}	The random part of L_B
$L(C)$	Feeder cable loss (dB)
$L_{empirical}$	Empirical value of the path loss in an urban setting (See Equation 5.35)
L_{ex}	Excess path loss (in addition to free space path loss)
L_F	Free space path loss

L_{floor}	Excess path loss due to passing through the floors of a building
$L_{gases}, L(G)$	Excess path loss due to the presence of atmospheric gases
L_I	Excess path loss in the final street (See Equation 5.39)
L_{ke}	Excess path loss due to diffraction by a knife edge
L_{max}	Maximum acceptable path loss
L_n	Excess path loss due to diffraction by the tops of n buildings
$L_{p.e.}$	Plane earth path loss
L_r	Receiver feeder line loss
$L_{rain}, L(R)$	Excess path loss due to rain
L_t	Transmitter feeder line loss
L_z	Excess path loss due to atmospheric gases for $\psi = 90°$ (zenith)
$L_{0.01}$	The value of L_{rain} when the rain rate $R_{rain} = R_{0.01}$
M	Number of cell phone clusters in a service area (i.e., the frequency reuse factor)
M_e	Mass of the earth
m_e	Mass of an electron (9.11×10^{-31} kilograms)
m_s	Mass of a satellite
N	Number of duplex channels available for trunking (in a cell phone system, the number of duplex channels in a cell)
N	Number of antennas in an array
N_o	Noise spectral density
N_T	Total number of available duplex channels (e.g., $N_T = n_c N$)
n	Refractive index
n_B	Number of buildings
n_c	Number of cells in a cluster
n_e	Electron number density (meters^{-3})
n_s	Settled value of n_B (See Equation 5.41)
n_T	Total electron content (See Equation 7.32)
P_C	In-cell power (Watts)
P_c	Power lost in conductors of microstrip patch cavity
P_d	Power radiated by a dipole antenna
P_d	Power lost in dielectric substrate of microstrip patch cavity
P_I	Interfering power (Watts)
P_{loss}	Power lost from a resonant cavity
P_N	Equivalent noise power at a receiver input
P_o	Time-averaged power dissipated by the Ohmic resistance of an antenna
P_{pe}	Power radiated by a patch edge

P_r	Radiated power (Watts)
P_s	Received signal power (Watts)
P_t	Transmitter power (Watts)
Pr [--]	Probability
$p(X)$	Probability density function (p.d.f.) of the random variable X
p_B	Probability of blocking a signal
p_{cell}	Probability of call completion averaged over the cell
p_e	Probability of call completion at the edge of a cell
p_s	Probability of call success
Q	Quality factor of a resonant cavity
$Q(M/\sigma)$	Complementary cumulative normal distribution function for p.d.f. with standard deviation σ
Q_o	Unloaded quality factor of a resonant cavity
q	Charge on a particle (Coulombs)
$R_{0.01}$	Rain rate exceeded 0.01% of the time
R_a	Distance of a satellite from the center of the earth
R_d	Radiation resistance of a dipole antenna
R_i	Input resistance of an amplifier
R_o	Ohmic resistance of an antenna
R_r	Radiation resistance of an antenna
R_{rain}	Rain rate (mm/hr)
R_s	Surface resistance of a conductor
R_{synch}	Distance of a geostationary orbit (GSO) satellite from the center of the earth
Re	Real part
r	Position vector in spherical coordinates (meters)
r	Range (i.e., distance between transmitting and receiving antennas) (meters)
r_B	Distance from base station to first building
r_c	Radius of a cell
r_e	Shadowing autocorrelation distance
r_{km}	Range in kilometers
r_l	Radius of a loop
r_m	Distance between sequential positions of a mobile communications device
r_R	Path length through the rain
r_w	Radius of a wire
r_1	Radius of first Fresnel zone (See Equation 5.27)
S	Poynting vector (Watts/meter²); time averaged power density carried by an electromagnetic wave
S	Power density (Watts/meter²)

S_I	Isotropic power density (Watts/meter²)
S_{max}	Maximum power density in a preferred direction
T	Temperature
T_A	Antenna temperature due to noise picked up from external sources
$T_{atmosphere}$	Temperature corresponding to noise from atmospheric gases
T_b	Temperature corresponding to the "big bang" noise
T_{cosmic}	Temperature corresponding to noise from extraterrestrial sources (aside from T_b)
T_e	Equivalent noise input temperature to account for noise generated by receiving amplifier
T_f	Feeder cable temperature
$T_{galactic}$	Temperature corresponding to noise coming from the galaxy outside our solar system
T_o	Ambient temperature (290°K)
T_{rain}	Noise temperature of the rain (260°K)
T_s	Temperature of large objects in the field of view of an antenna
t	Time (seconds)
U	Traffic intensity in erlangs
u	Fraction of radiation beam of a receiving antenna filled by the sun or the moon
V_A	Noise voltage due to noise picked up by antenna from external sources
V_F	Voltage to account for noise generated by receiver amplifier
V_i	Voltage across input terminals
V_N	Noise voltage
V_o	Noise voltage due to Ohmic resistance of an antenna
v_g	Group velocity
v_s	Speed of a satellite
W	Energy (Joules)
W_s	Energy per symbol in digital communications
w	Thickness of dielectric substrate in microstrip patch antennas Cartesian coordinate (meters)
Y	Admittance (Siemens)
y	Cartesian coordinate (meters)
Z	Impedance (Ohms)
Z_o	Wave impedance of free space; 377 Ohms
z	Axial coordinate (meters)

Greek Alphabet

α, Alpha	Incremental phase advance between adjacent antennas in a colinear array
α_c	Attenuation constant
β, Beta	Phase constant
Γ, Gamma	Average value of the signal-to-noise ratio
γ	Signal-to-noise ratio
γ_{min}	Minimum acceptable value of the signal-to-noise ratio
γ_R	Specific rain attenuation (dB/km)
Δ, Delta	Distance from dipole to corner in a corner-reflector antenna
Δ_a	Difference in phase angle
Δf	Bandwidth
Δ_h	Difference in height
Δz	Length of a Hertzian dipole
$\tan \delta$	Loss tangent
$\delta(\mathbf{r} - \mathbf{r_0})$	Dirac delta function
δf	Doppler frequency shift
δf_m	Maximum Doppler frequency shift
δ_s	Skin depth
ε, Epsilon	Permittivity
$\varepsilon^{'}$	Real part of the permittivity
$\varepsilon^{''}$	Imaginary part of the permittivity
ε_0	Permittivity of free space
ε_r	Relative permittivity
ε_{xy}	Permittivity tensor of a magnetized plasma
ζ, Zeta	Angle between direct path from base station antenna to the edge of the last building, and the horizontal
η, Eta	Antenna efficiency based on Ohmic losses
η_{ap}	Aperture efficiency
θ, Theta	Polar angle coordinate
θ_B	Brewster angle
θ_i	Angle of incidence
θ_n	Phase of electric field of signal arriving along path n
θ_r	Angle of reflection
θ_t	Angle of transmission
κ, Kappa	Flat-edge model parameter (See Equation 5.37)
κ_1, κ_2	Terms in permittivity tensor of magnetized plasma
Λ, Lambda	Largest antenna dimension
λ	Wavelength

λ_o	Free space wavelength
$\underline{\lambda}$	Dimensionless wavelength normalized to 1 meter
μ, Mu	Permeability
μ	Mean value of a random variable
μ_o	Permeability of free space
μ_r	Relative permeability
ν, Nu	Fresnel diffraction parameter
Π, Pi	Power spectral density
ξ, Xi	Phase angle
ρ, Rho	Charge density (Coulombs/meter^3)
$\rho(n)$	Vector from nth antenna to far field observer
ρ_{FN}	Distance along the ground from the base station antenna in which nulls in the radiation pattern occur
ρ_{max}	Distance along the ground from the base station antenna to the maximum in the radiation pattern
$\rho_s(r_m)$	Autocorrelation function
Σ, Sigma	Area of a loop antenna
σ	Standard deviation
σ_B	Standard deviation of building penetration loss
σ_c	Conductivity (Siemens/meter)
σ_{cu}	Conductivity of copper (5.8×10^7 Siemens/meter)
σ_h	Standard deviation for the height of a rough surface
σ_L	Standard deviation of shadowing
τ, Tau	Time shift
τ_e	Time for a satellite to complete an orbit of the earth
τ_s	Time duration of a symbol
Φ, Phi	Scalar potential
ϕ	Azimuthal angle coordinate
χ, Chi	Phase angle
ψ, Psi	Elevation angle
Ω, Omega	Ratio of in-cell signal power to interfering power
ω	Angular frequency in radians/second
ω_c	Electron cyclotron frequency (See Equation 7.39)
ω_p	Plasma frequency (See Equation 7.16)
ω_s	Angular speed of a satellite

Index

Milton Keynes UK
Ingram Content Group UK Ltd.
UKHW031128141024
449569UK00006B/349

9 781439 878972